Topology of 4-Manifolds

PRINCETON MATHEMATICAL SERIES

Editors: LUIS A. CAFFARELLI, JOHN N. MATHER, JOHN D. MILNOR,
AND ELIAS M. STEIN

Topology of 4-Manifolds

Michael H. Freedman and Frank Quinn

Princeton University Press
Princeton, New Jersey
1990

Library of Congress Cataloging-in-Publication Data

Freedman, Michael H., 1951–
  Topology of 4-manifolds / Michael H. Freedman and Frank Quinn.
    p.   cm. — (Princeton mathematical series ; 39)
  Includes bibliographical references.
  ISBN 0-691-08577-3 (alk. paper)
    1. Four-manifolds (Topology)   I. Quinn, F. (Frank), 1946–
II. Title.   III. Title: Topology of four-manifolds.   IV. Series.
QA613.2.F75 1990
514′.3—dc20   89-48809

Printed in the United States of America by Princeton University Press,
Princeton, New Jersey
10 9 8 7 6 5 4 3 2 1

Contents

Topology of 4-Manifolds

Introduction

The study of the topology of manifolds has turned out to be dependent on dimension in a curious way. Manifolds of dimension 2 are a classical subject, and have been largely understood for half a century. Historically the next major progress was in dimension 3; Moise in 1952 reduced the topological theory to the piecewise linear (or equivalently smooth) category, and Papakyriakopoulos in 1957 established the beginnings of embedded surface theory. But then dimension 4 was skipped over, and higher dimensions (≥ 5) were developed next. The two key events were the development of the methods of smooth and PL handlebody theory by Smale in the late 1950s, and the extension to topological manifolds by Kirby and Siebenmann in 1969. Dimension 4 was last, with the corresponding results published by the authors in 1982.

From a technical point of view it is the behavior of 2-dimensional disks which separates the dimensions in this way: the generic map of a 2-disk in a 3-manifold has 1-dimensional selfintersections; in a 4-manifold the intersections are isolated points; and in dimensions 5 and above 2-disks are generically embedded. Progress in each realm has been based on understanding and use of these disks. Papakyriakopoulos' breakthrough in dimension 3 was a proof of the "Dehn lemma" for embedding 2-disks in 3-manifolds. In dimension 5 and higher 2-disks are used in the "Whitney trick," a central ingredient of geometric topology. Also the embedability of 2-disks is the key to a characterization of these manifolds†. Similarly the major technical results presented here are embedding theorems for 2-disks in 4-manifolds.

New phenomena are also encountered in dimension 4, which is one reason the development took a decade longer. In other dimensions the principal methods (handlebody theory or embedded surface theory) proceed uniformly in the smooth, PL or topological categories, once the basic machinery has been set up. Differences between the categories can for the most part be understood in terms of the associated bundle theories. In

---

†If $X \times \mathbf{R}^k$ is a manifold for some k, then X itself is a manifold of dimension 5 or greater if and only if maps of a 2-disk into X can be arbitrarily closely approximated by embeddings; Quinn [2].

dimension 4 qualitatively different—and more subtle—phenomena occur
in the smooth category, revealed particularly by the work of S. Donaldson. Here we deal with the topological theory, where the results parallel
those of higher dimensions. The proofs however are substantially more
complicated: the smooth analogs are known to be false by Donaldson's
work, so the proofs must require enough purely topological complication
to avoid proving these analogs.

Our objective is to present the topological theory of 4-manifolds. As
mentioned above this dimension has its own unique flavor and complication. However when possible, particularly in the structure theory of
Part II, comparisons are made with other dimensions. This material
should not be viewed as special and isolated, but as part of the general
development of manifold theory which is one of the major mathematical
achievements of our century.

The topics and proofs reflect the prejudices of the second author,
who did most of the writing. In particular in Part I the properties
of immersed surfaces are extensively developed, and the "calculus of
links"—devised by R. Kirby and decisively employed by the first author in many of the original proofs—is not used. In Part II we note
the omission of pseudoisotopies and isotopies, controlled ends, the more
detailed applications of surgery, and any substantial discussion of the
differential-geometric methods used in the smooth category.

We would like to express our appreciation to F. D. Ancel, R. Gompf,
Richard Stong, and M. Yamasaki for their comments on earlier versions.
We take this opportunity to acknowledge the many contributions (often
unpublished, but freely shared) that R. D. Edwards has made to this
subject. Finally, we have benefited from support from the National
Science Foundation and many other institutions.

Outline

The material divides naturally into two parts; embedding theorems,
and applications of these to the structure of manifolds.

The embedding theorems for the most part provide criteria for the
existence of topological locally flat embeddings of 2-disks in a 4-manifold.
These theorems are specialized to dimension 4, and presented in detail in
Part I (Chapters 1–6). The development is detailed and elementary, and
should be accessible to most graduate students. Exercises are provided
since a hands-on approach is the best way to learn to see things in 4-
D. References and historical notes in this part are collected in the final
section of each chapter.

Much of the material of Part I is specialized to the proof of the embedding theorems, and is not directly used in Part II. For a first reading

we suggest the general material of Chapter 1, the statement of theorem 5.1, and selections from Part II (eg. Chapters 10 and 11). For a thorough understanding of Part II, in addition to Chapter 1 the first few sections of Chapter 2 and the remaining statements in Chapter 5 seem to be sufficient.

Applications to the structure of topological 4-manifolds are given in Part II, Chapters 7–12. This is less self-contained than Part I since some of the results rely on machinery developed in high-dimensional topology. Where feasible we have sketched this machinery, or given arguments adapted to this dimension. We have also surveyed analogous material from higher dimensions, to make plain the similarities and pecularities of the 4-dimensional results.

The most important conclusion about embeddings of a single disk is:

Embedding theorem. *Suppose* $A: D^2 \to M^4$ *is an immersion with boundary going to boundary, and* $B: S^2 \to M$ *is a framed immersion with algebraic selfintersection* $\mu(B) = 0$ *and intersection* $\lambda(A, B) = 1$.

If the fundamental group of M *is poly-(finite or cyclic) then there is a topologically framed embedded disk in* M *with the same framed boundary as* A.

The hypothesis of the existence of the sphere B is purely algebraic-topological (equivalent to a class in $H_2(M; \mathbf{Z}[\pi_1 M])$ with certain properties), and is satisfied in most cases of interest. The fundamental group hypothesis is much more restrictive, and the principal open question in the subject is whether the theorem holds without it (this seems unlikely).

A group is poly-(finite or cyclic) if there is a finite ascending sequence of subgroups, each normal in the next, so that successive quotients are either finite or infinite cyclic. In fact the proof and some simple observations show that the theorem is valid for fundamental groups in the following class: the smallest class of groups containing finite and cyclic groups, which is closed under direct limits, subgroups, quotients, and group extensions. This class contains some groups which are not poly-(finite or cyclic), but the more concrete term has been retained because it is descriptive and seems to cover the interesting cases. "Large" groups, for instance ones containing a free group with two generators, are not included.

The other main result is the *controlled embedding theorem*, which gives control over simultaneous embeddings of many disks. The setting is a map $M \to X$, from a 4-manifold to a metric space. Roughly speaking we show that if $\epsilon > 0$ then there is $\delta > 0$ so that a collection of immersed disks and dual spheres whose images in X have diameters less than δ, can be replaced by embeddings whose images in X have diameter less than ϵ. This refinement is used in the proof of the annulus theorem and many results about local topological behavior.

The first chapter introduces the basic tools for manipulating maps of surfaces to 4-manifolds, including immersed Whitney moves and transverse spheres. Almost no background is assumed. The material in this chapter is generally useful in studying 4-manifolds, while the later material enters primarily in the proof of the disk embedding theorem.

The second chapter deals with capped gropes. These are the principal technical innovation discovered since the original publications on the subject. The idea is that "more subtle" types of selfintersections of a disk can be obtained by constructing complicated 2-complexes which have the same regular neighborhood as a disk, and restricting the intersections permitted in this 2-complex. With the hypotheses of the embedding theorem (but arbitrary fundamental group) a properly immersed capped grope can be constructed. To find a disk we need an immersed capped grope whose image has fundamental group mapping trivially to the fundamental group of the manifold. We are able to reduce the fundamental group enough to avoid poly-(finite or cyclic) groups, but not for example free groups. This is where the fundamental group hypothesis enters in the main theorem.

The third chapter concerns capped towers. Like capped gropes these are 2-complexes with the same regular neighborhood as a disk. They have a more complex internal structure than a grope, exploiting the hypothesis that the image maps trivially on fundamental group into the whole manifold. A replication theorem allows an infinite construction, of "convergent infinite towers" in a manifold. These are singular approximations to neighborhoods of embedded disks.

Chapter four develops some decomposition space material, in the tradition of M. Brown and R. H. Bing. The objective of these techniques is to determine when a map can be approximated by a homeomorphism. Here they are used to identify the singular objects constructed in Chapter three as often being topological framed embedded disks.

The fifth chapter gives formal statements of the embedding theorem and its variants, and assembles the proofs. In addition to the poly-(finite or cyclic) and controlled results referred to above, there is a version in which a π_1-nullity hypothesis on the image of the data replaces the condition on π_1 of the manifold.

The sixth chapter deals with an embedding theorem in which the manifold is allowed to change slightly (by s-cobordism). This flexibility allows a slight weakening of the π_1-null hypothesis in the general embedding theorem.

Part II develops consequences of the embedding theorems for the structure of manifolds. Roughly, the first three chapters are devoted to developing the basic machinery previously known to be available in other dimensions. Chapters 10 and 11 then give specific structural results, and Chapter 12 describes attempts to extend the class of fundamental groups for which the theorem is known.

Chapter 7 gives variations on the 5-dimensional h-cobordism theorem. The classical form, that an h-cobordism with good fundamental group and vanishing torsion has a product structure, has the 4-dimensional Poincaré conjecture as an immediate consequence. There is a controlled version which is even more extensively used, and a proper version is obtained as a corollary of this.

Smooth structures on 4-manifolds are studied in Chapter 8. The main conclusions are existence and uniqueness theorems, though they are weaker than results in other dimensions because they are constrained by Donaldson's nonexistence results. Both Donaldson's theory and the higher dimensional situation are reviewed.

The ninth chapter is a technical one, establishing the existence of handlebody structures, normal bundles, and transversal approximations. In each case the result completes the picture, filling the gap left by developments in other dimensions.

Chapter 10 addresses the classification of 4-manifolds. Closed manifolds with trivial or infinite cyclic fundamental groups are classified in terms of bilinear forms over the group ring. The heart of the chapter is an existence and uniqueness theorem for certain embeddings of bounded 4-manifolds, in 4-manifolds with good fundamental group.

In Chapter 11 "surgery" for 4-manifolds is developed and some of the consequences are surveyed. Some of the classification theorems for aspherical manifolds extend to this dimension. Trivial knots in S^4 and certain slice knots in S^3 are characterized. Tame ends (with good fundamental group) of 4-manifolds are classified. There are also applications to the structure of isolated fixed points of actions of finite groups on 4-manifolds.

Finally in Chapter 12 we present some approaches to the "main problem": is the embedding theorem valid without the poly-(finite or cyclic) π_1 hypothesis? There are straightforward reformulations in terms of slices for certain families of links, and less direct ones involving transversality of Poincaré spaces, and actions of free groups on the 3-sphere.

PART I

Embeddings of Disks

CHAPTER 1

Basic tools

In this chapter the principal methods for manipulation of immersions of surfaces are described. The raw material is provided by the immersion lemma 1.2, which is assumed as an axiom in Part I. This and the twisting operation of 1.3 are used to construct (immersed) Whitney disks in 1.4. These disks are used to define Whitney moves, which are one of the principal geometric operations in topology of any dimension. Finger moves and connected sums are developed in sections 1.5 and 1.8 as ways to remove intersections, especially with Whitney disks. Intersection numbers, described in 1.7, give algebraic conditions under which the geometric operations can be used. One of the central ingredients of later constructions, transverse spheres, are introduced in 1.9. Finally section 1.10 describes "scorecards," a bookkeeping device useful for recording intersections among families of surfaces.

1.1 Pictures in 4-space

We picture 4-space as the familiar 3 space dimensions, and time. A picture of a 3-dimensional object therefore represents something in a slice ("the present") of $\mathbf{R}^4$. The strategy will be to compress as much as possible of the activity into this slice. Several conventions will be used in the pictures. Solid lines will represent things in the present time. Dotted lines represent things in the future, dashed lines are in the past. We suggest that the reader draw over the dotted lines in color (eg. blue = future, red = past). Color is much more effective in suggesting another dimension, but we cannot print the pictures in color. A line in the present which passes behind something is drawn thinner (since we cannot use the standard convention of drawing it dashed). Finally, we sometimes use a convention to avoid drawing things in the past or future. If a surface is drawn as a line, it is understood that it extends into the past and future simply as product with $\mathbf{R}$.

In the first example a plane A is shown entirely in the present. B starts in the past, intersects the present in a line, and extends into the future. This is shown with and without the time extension convention.

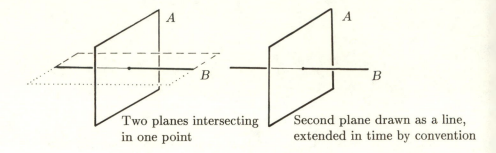

Two planes intersecting   Second plane drawn as a line,
in one point   extended in time by convention

Since A lies entirely in the present the intersection with B is easy to see: it is the single point of intersection with the line of B lying in the present.

Exercise   Suppose c is an arc in a plane, and a neighborhood of the plane is parameterized as $\mathbf{R}^2 \times \{0\} \subset \mathbf{R}^2 \times \mathbf{R}^2$. The *linking annulus* of the arc is the annulus $c \times S^1 \subset \mathbf{R}^2 \times \mathbf{R}^2$. (1) Let c be the line of intersection of the plane B with the present, and draw the linking annulus. (2) Draw the *linking torus* of an intersection point; if a neighborhood of the point is parameterized as $\mathbf{R}^2 \times \mathbf{R}^2$, then the torus is $S^1 \times S^1$. In the picture above this torus intersects the present in two circles, and extends to a copy of $S^1 \times D^1$ in both the past and the future. ▯

1.2 Framed immersions

These provide models for images of surfaces mapped into 4-manifolds. Suppose N is a surface, and let D, E be disjoint copies of $\mathbf{R}^2$ in N. Form $N \times \mathbf{R}^2$, and identify $D \times \mathbf{R}^2$ with $E \times \mathbf{R}^2$ by the isomorphism which switches the coordinates. This is a *plumbing* of $N \times \mathbf{R}^2$ with itself. The identification space obtained by some finite number of disjoint plumbings is a smooth manifold. The image of N in this is embedded except at isolated double points, which have neighborhoods isomorphic to the picture drawn in 1.1.

A *framed immersion* of N into M^4 is an isomorphism of an open set in M with such a plumbing. A framed immersion defines a map $N \to M$ with isolated double points, and to simplify notation we often refer to this map as the framed immersion. It is quite important, however, that a framed immersion specifies a neighborhood of the image as well as the image itself. When it is necessary to be precise we say that the map *extends* to a framed immersion. Note that the plumbing has boundary $(\partial N) \times \mathbf{R}^2$, and the boundary of an open set in M is the intersection with ∂M. Therefore immersions under this definition are automatically "proper" in the sense that they take ∂N to ∂M.

Consider D^2 as the unit disk in the complex numbers. If $r\colon N \to S^1$ is a map we can define an automorphism of $N \times D^2$ by multiplying by $r(n)$ in $\{n\} \times D^2$. Two framed immersions *differ by rotations* if one is obtained from the other by composing with such an automorphism of $N \times D^2$. Similarly we can let $-1 \in S^0$ act on $\mathbf{R}^2$ by $\left[\begin{smallmatrix} 1 & 0 \\ 0 & -1 \end{smallmatrix}\right]$, so a map $r\colon N \to S^0 = \{\pm 1\}$ defines an automorphism of $N \times \mathbf{R}^2$. Two immersions *differ by sign* if they differ by an automorphism of this form.

Immersion lemma. *Suppose $f\colon N^2 \to M^4$ is a map.*

(1) *If $N = D^2$ or $\mathbf{R} \times I$ and f extends to a framed immersion in a neighborhood of ∂N then there is a homotopy of f rel ∂N to a map which extends to a framed immersion, which near ∂N differs from the given one by rotations ($N = D^2$) or sign ($N = \mathbf{R} \times I$).*

(2) *If N is the union of open sets $U \cup V$, and the restrictions of f extend to framed immersions which agree on $U \cap V \times \mathbf{R}^2$, then f can be changed by isotopy on U rel $U \cap V$ so that the result extends to a framed immersion of the union.*

Note that any surface can be decomposed into disks and ribbons, so (1) can be used to build up immersions of arbitrary surfaces. The signs and rotations in the normal direction define a normal vector bundle for the image.

The point of (2) is that U and V might individually be good, but the intersections between them not come from plumbings. When this *is* the case (ie. f defines an immersion—in the sense above—of $U \cup V$) then the pieces are then said to be "in general position," or "transverse."

This lemma provides the starting data for most of our constructions. In the smooth and PL categories it is one of the simplest cases of transversality and general position, and can be found in many text-books. We will not give a proof here. The topological version is proved in section 9.4C, as a consequence of the smoothing theorem proved in chapter 8. It is a deep fact, and indeed depends on the whole development in the smooth or PL case.

The image of $N \times D^2$ in a plumbing of $N \times \mathbf{R}^2$ is a regular neighborhood of the image of N. If each component of N has nonempty boundary then this regular neighborhood is also a regular neighborhood of a graph; choose a graph in N which has N as a regular neighborhood. At each intersection point add two embedded arcs, one in each sheet, joining the intersection point to the graph. Then the regular neighborhood of the image is isotopic to a regular neighborhood of this enlarged graph.

As a consequence of this description of the regular neighborhood we have a useful description of the fundamental group of the image: it is free, and if N is connected it is generated by loops passing through at most one intersection point. This is based on the assumption that components of N have nonempty boundary, but the π_1 conclusion remains true if the closed components are spheres.

1.3 Twisting

The twisting operation changes an immersion so as to change the framing of the boundary. It can often be used to undo the rotations encountered in the immersion lemma.

Consider a standard $D^2 \subset D^4$. Cut out a square in the interior, and replace it by two pieces; a strip twisted about one edge, and a piece obtained by filling in straight lines between points on the opposite edge and corresponding points on the twisted strip.

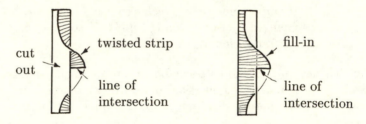

As described, this object is completely in the present. Both the twisted strip and the fill-in piece have a line of intersection with the D^2. Perturb the interior of the twisted strip into the past, and the interior of the fill-in piece slightly into the future. The result is an immersion which has a single selfintersection point. Near the boundary it differs from the original by rotations, specifically two full twists; ± 2 in $\pi_1 S^1 \simeq \mathbf{Z}$. This can be seen directly—one twist is evident in the twisted strip, the other comes from the shifting into the past and future—or it can be deduced from the presence of the single selfintersection point using the formula relating intersections and Euler numbers (1.7).

There is a technically more important "boundary twist" operation, but before considering that we give a corollary of interior twisting.

1.3A Corollary. *A map $f: S^2 \to M^4$ is homotopic to a framed immersion if and only if $\omega_2(f) = 0$.*

Here the second Stiefel-Whitney class $\omega_2 \in H^2(M; \mathbf{Z}/2)$ is thought of as a homomorphism $\omega_2: H_2(M; \mathbf{Z}/2) \to \mathbf{Z}/2$, and $\omega_2(f)$ denotes the value of this on the image of the orientation class of S^2.

Proof: Arrange, by homotopy, that f intersects a ball in M in a standard disk $D^2 \subset D^4$. Let M_0 denote the manifold obtained by deleting the interior of the ball of radius $1/2$, then f defines a map $f_0 \colon D^2 \to M_0$ which is an immersion near the boundary. Apply the immersion lemma to find a homotopic immersion $f_0{}'$ which differs from the given one by rotations near the boundary. The number of full twists in the rotation function is equal mod 2 to $\omega_2(f)$, so $\omega_2(f) = 0$ if and only there are an even number of full twists. In this case we can apply the twisting operation in the interior of the disk to obtain an immersion with null-homotopic rotation function. A nullhomotopy can be used to define a framing of the union of the standard disk and the twisted $f_0{}'$, giving a framed immersion.

To see the other direction (and the fact about $\omega_2(f)$) requires some background we will not review in detail. Briefly, when the rotations are nontrivial the standard disk and $f_0{}'$ fit together to define an "unframed" immersion of S^2. The normal bundle of this immersion has the rotation function as clutching function, so the degree of the rotation function determines the bundle as an element of $\pi_2(\mathbf{BO}(2)) \cong \mathbf{Z}$. The degree mod two therefore determines the stable class in $\pi_2(\mathbf{BO}) \cong \mathbf{Z}/2$. But this stable class depends only on the homotopy class of f (and the tangent bundle of M) and is determined by $\omega_2(f)$. Therefore if $\omega_2(f) \neq 0$, f cannot be represented by an immersion with trivial normal bundle. ∎

Now we turn to "boundary twisting." For this we need the notion of a surface B being immersed with part of its boundary on another surface A. The model for such a thing is built from two surfaces A and B, an embedded collection of arcs and circles $c \subset \partial B$, and an embedding $c \times [-1, 1] \to A$. Locally the model looks like:

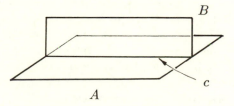

Explicitly, choose a collar $c \times [0, 1] \subset B$, then there are subsets $c \times [0, 1] \times [-1, 1] \subset B \times [-1, 1]$ and $c \times [-1, 1] \times [-1, 1] \subset A \times [-1, 1]$. We identify the first with the image in the second, by the map which switches the last two coordinates. This defines a 3-manifold neighborhood of the union $B \cup_c A$, and a 4-manifold is obtained by product with I. Immersions of this object are defined by open embeddings of plumbings disjoint from the juncture c.

Now consider the model for a disk embedded in D^4 with part of its boundary on the standard 2-disk (ie. the local model pictured above).

Cut out a square in B, with one edge on A. Replace this by a twisted strip, twisted about the edge on A, and a fill-in piece as above:

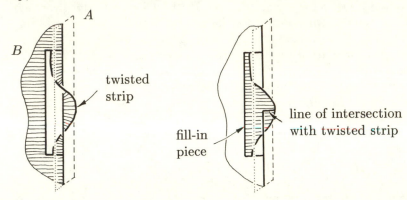

The result is a map B' which is a framed immersion of $D^2 - c$, and near A can be parameterized as a standard immersion of a collar on c attached to A. The framings of these two immersions differ by rotation by one full twist near c. Note that B' has a new intersection point with A.

1.3B Corollary. *Suppose $c \subset S^1$ is nonempty, $c \times [-1, 1] \to S$ is an embedding in a surface, $A: S \cup_c (S^1 \times [0, 1)) \to M^4$ is a framed immersion, and $f: D^2 \to M$ is a nullhomotopy of the restriction of A to S^1. Then there is a framed immersion of $S \cup_c D^2$ which agrees with A on $S \cup_c S^1$ and is homotopic rel boundary to f on D^2.*

Proof: Denote by M_0 the complement of a small regular neighborhood of the image of c. The immersion A defines an immersion A_0 of the complement of a neighborhood of c in S, disjoint union with $S^1 \times [0, 1)$, to M_0. f extends the latter to a map of a disk. Applying the immersion lemma gives an immersion $B: D^2 \to M_0$ homotopic to f which near S^1 differs from the given immersion by rotations. Replace c, and apply the boundary twisting procedure near a point in c to cancel the rotations. The result is the desired framed immersion. ∎

Exercise Suppose T is a surface, $c \subset \partial T$, and $c \times [-1, 1] \to S$ is an embedding. Let $A: S \cup_c (\partial T \times [0, 1)) \to M$ be a framed immersion which extends to a map of $S \cup_c T$ to M. Give a criterion generalizing the two corollaries for the existence of a framed immersion of $S \cup_c T$ which agrees with A on $S \cup_c \partial T$ and is homotopic rel boundary to the given map on T. For this use the fact that a map of a surface can be approximated by an embedding on a neighborhood of an arc, with the image of the arc disjoint from all other surfaces. Recall also that surfaces can be reduced to disks by cutting along arcs and circles. □

1.4 Whitney disks

Suppose immersed surfaces A, B intersect in two points, and there is an embedded 2-disk W with boundary on the union of the images, as shown:

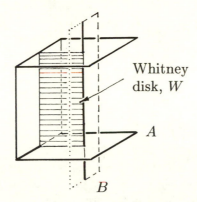

This picture is the standard model for a neighborhood of a *Whitney disk*. The boundary of this disk (a union of two arcs, one on each surface) is called a *Whitney circle* for the intersection points. Finally we say that a pair of intersection points *have opposite sign* if they have a Whitney circle; if there are embedded arcs joining them on A and B with a neighborhood isomorphic to a neighborhood in the model.

When A, B, and M have appropriate orientation data, and A and B are connected, then this definition of "opposite sign" agrees with the definition of signs for intersection numbers (see 1.7). If a half twist is inserted in A or B in the picture above then the intersection points have the "same sign," and the circle cannot appear as the boundary of a Whitney disk.

An application of Corollary 1.3B in the previous section yields:

Lemma. (Immersed Whitney disks) *Suppose c is a Whitney circle in $A \cup B$ which is contractible in M. Then there is an immersed Whitney disk with boundary c.*

As above "immersed" means a map whose image has a neighborhood obtained by introducing plumbings into the standard model. W is allowed to intersect A and B, as well as itself.

Let W denote the model Whitney disk for a pair of intersection points $A \cap B$. There is an ambient isotopy (the *Whitney move*) supported in a neighborhood of W which moves A to a surface A' disjoint from B.

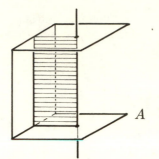

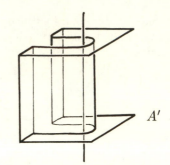

A' can be described as constructed from A by cutting out a neighborhood of the arc on A, glueing in two parallel copies of the Whitney disk, and a parallel of a neighborhood of the arc on B.

Most often we will only be able to find an immersed Whitney disk for a pair of intersections. In this case the Whitney move produces a homotopy rather than an isotopy. The description of A' as constructed from parallel copies of W makes it clear that A' will intersect everything that intersects W.

Exercise (1) Suppose W is a framed immersed Whitney disk on $A \cap B$. Count the number of new intersections of A' with B and itself, in terms of intersections and selfintersections of W. (2) Suppose $A\colon S^2 \to M^4$ is an immersion with selfintersections arranged in pairs, with immersed Whitney disks. Let A' be a parallel copy of A. Find all intersections of A and A', and show that they can be arranged in pairs with immersed Whitney disks (essentially parallel copies of the original ones). □

1.5 Finger moves

Suppose there are surfaces A, B, and W in M, part of the boundary of W lies on B, and there is an embedded arc from an intersection point $A \cap W$ to this boundary, as shown. Then we can *push A off W through B* (along the arc). This gives a surface A' with one fewer intersection with W, but two new intersections with B.

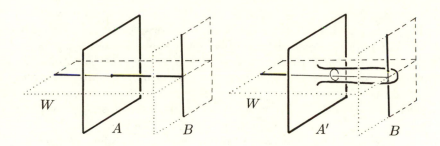

This is called a "finger move" because one imagines producing the deformation by poking a finger through A. Technically a disk in A around the intersection point is replaced by the linking annulus of the arc, and a disk at the other end which misses B. This can be thought of as the inverse of a Whitney move, since the new pair of intersection points $A \cap B$ have a canonical Whitney disk, whose Whitney move pushes A' back to A.

Exercise   Draw this Whitney disk. □

An important use of finger moves is to push surfaces off Whitney disks. This fits together with several other moves to give a unit we formalize as a lemma.

Immersed Whitney move. *Suppose $A, B \to M$ are framed immersed surfaces, with contractible Whitney circles for all intersections. Then there are homotopic immersions $A', B' \to M$ with the same framed boundaries and $A' \cap B' = \phi$.*

Proof: Apply the lemma of 1.4 to find immersed Whitney disks W_i with the given Whitney circles as boundaries. Choose disjoint embedded paths from points of $B \cap W_i$ to the edge of W_i lying on B. Do finger moves, pushing B along these paths off W_i, through B. This gives B' disjoint from the interior of W_i, but with more selfintersections. Now use the Whitney move to push A across W_i. The result is A' such that $A' \cap B' = \phi$.

Notice that A' also has new selfintersections, which arise during the Whitney move from the intersections $A \cap W_i$ and selfintersections of W_i.

Exercise   Count the new selfintersections of A' and B', in terms of intersections with W. □

1.6 Regular homotopy

This is the analog of homotopy for immersions. A (general position, rel boundary) *regular homotopy* between framed immersions is a series of modifications, each of which is either (1) an ambient isotopy of the image rel boundary, or (2) a Whitney move, or (3) a finger move. The usual definition of regular homotopy is somewhat weaker; a homotopy through local embeddings. A generic smooth regular homotopy can easily be seen to decompose into a sequence of moves as in the current definition. The same is true of topological regular homotopies, but it is much harder to show. We avoid the difficult theorem simply by using the sequence-of-moves description as the definition.

For example the immersed Whitney move above gives regular homotopies of A, B to A', B'. In contrast, the twisting operation gives homotopic immersions which are not regularly homotopic; the modifications permitted in a regular homotopy do not change the framing of the boundary, whereas twisting does.

We formalize this example by defining a *cusp homotopy* of $D^2 \to D^4$. Begin with the standard embedding, then perform a twist about some arc. Use radial contraction to define a homotopy from the twisted immersion to the original. Note the twisted immersion has a single self-intersection point, which disappears during the homotopy. The picture shows a cross-section which illustrates this.

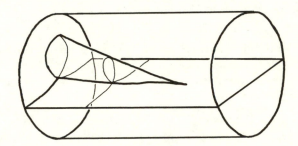

There is an alternative description more directly based on the picture. Begin with an immersion $D^1 \to D^2$ with a single selfintersection, as on the left in the picture. Define a homotopy to the standard embedding by pulling out the kink. Cross with $D^1 \subset D^2$ to get a homotopy of D^2 in D^4. The problem with this is that where the D^1 has a selfintersection point, the corresponding $D^2 \to D^4$ has a line of selfintersections. Resolve these lines to single points by perturbing part of the line into the past, and part into the future.

Proposition. *A homotopy between immersions of a surface in a 4-manifold is homotopic to a composition of homotopies, each of which is a regular homotopy or a cusp homotopy in some ball, or the inverse of a cusp homotopy.*

In the smooth and PL categories this is a simple consequence of singularity theory. The topological version—like the topological version of 1.2—requires most of the theory we are developing.

1.7 Intersection numbers

Intersection and selfintersection numbers give algebraic access to some of the intersection structure of a framed immersion. In what follows A and B are framed immersions of unions of disks or spheres, in a 4-manifold M.

First we characterize certain important special values (0 or 1) for these numbers. The statement that the intersection number $\lambda(A, B)$ is zero is equivalent to saying that the intersections of A and B can be arranged in pairs, with opposite signs and contractible Whitney circles. Vanishing of the selfintersection number $\mu(A)$ is equivalent to a similar pairing of the selfintersections of A. Finally intersection 1 means that all but one intersection can be so paired.

These special cases are sufficient for the development in Part I. The general situation is described for reference in Part II. See Wall [**1**, §5] for more detail.

Generally intersection "numbers" lie in the group ring of the fundamental group $\mathbf{Z}[\pi_1 M]$, and require a little more data to define. Choose paths from the basepoint of M to the images of A and B. Choose orientations for A and B, and an orientation for M at the basepoint. The intersection number $\lambda(A, B)$ is the element of $\mathbf{Z}[\pi_1 M]$ obtained by adding up elements σg for each intersection point, where σ is 1 or -1, and $g \in \pi_1 M$. The element g is represented by a loop going from the basepoint to A, then in A to the intersection point, and back to the basepoint through B. Since A and B are simply connected (D^2 or S^2) this does not depend on the choice of paths in A and B.

To define σ for an intersection point, note that the orientations of A and B define an orientation of M at the intersection point. The chosen orentation at the basepoint can be transported along the path to A and through A to the intersection point. If these two orientations agree $\sigma = 1$, otherwise $\sigma = -1$.

Note that interchanging A and B reverses the direction in the loop, and changes the sign by $\omega_1(g)$, where $\omega_1 \colon \pi_1 M \to \mathbf{Z}/2$ detects whether or not a loop is orientable. Define an involution $\bar{\omega}$ on $\mathbf{Z}[\pi_1 M]$ by

$$\bar{\omega}\left(\sum n_g g\right) = \sum_g n_g \omega(g) g^{-1},$$

then the relation is $\lambda(B, A) = \bar{\omega}(\lambda(A, B))$.

Selfintersection numbers $\mu(A)$ are defined similarly, but lie in the quotient $\mathbf{Z}[\pi_1]/\{a-\bar{\omega}(a)\}$. The indeterminacy results from the fact that at a selfintersection there is no natural way to choose a first and second sheet; the quotient is obtained simply by dividing out differences between the two choices.

Intersection and selfintersections of a framed immersion A are related by the formula $\lambda(A, A') = \mu(A) + \bar{\omega}(\mu(A))$. Here A' is a parallel copy of A, and the sum is well-defined in $\mathbf{Z}[\pi_1]$ even though μ lies in a quotient. This is a generalization of the "evenness" of intersection numbers over $\mathbf{Z}$: if M is orientable the augmentation $\mathbf{Z}[\pi_1 M] \to \mathbf{Z}$ is invariant under $\bar{\omega}$, so the image in the integers satisfies $\lambda = 2\mu$.

In the non-framed case there is a correction to this formula. Suppose A is an immersed sphere whose normal bundle has Euler number χ, or is an immersed disk with a fixed framing of the boundary which has rotation number χ with respect to the normal bundle of the immersion. Let A' be the image of a section of the normal bundle, which in the disk case is parallel to the fixed framing on the boundary. Then the formula is $\lambda(A, A') = \mu(A) + \bar{\omega}(\mu(A)) + \chi$. Note the correction term is an integer, considered as an element of $\mathbf{Z}[\{1\}] \subset \mathbf{Z}[\pi_1]$.

In detail, the normal bundle E of such an immersion is obtained by glueing together trivial bundles on the two hemispheres of S^2 by rotations along the equatorial S^1. The degree of the rotation map $S^1 \to S^1$ is the Euler number $\chi(E)$ of the bundle, and also appears as the intersection number of the 0-section of the bundle with any other section in general position. Now suppose A is an unframed immersion of S^2 and A' is a "parallel copy;" the image of a section of the normal bundle. Then A intersects A' in two points for each selfintersection of A, and additional points coming from the intersections in the bundle. Figuring out the signs and loops associated to these points gives the formula.

Proposition. *Intersection numbers, and reduced selfintersection numbers in $\mathbf{Z}[\pi_1]/(\{a - \bar{\omega}a\} + \mathbf{Z}[\{1\}])$ are invariant under homotopy rel boundary. The $\mathbf{Z}[\{1\}]$ component of the selfintersection number is invariant under regular homotopy, and conversely two immersions of a sphere or disk which are homotopic rel boundary, and have the same framed boundary, are regularly homotopic rel boundary if and only if the $\mathbf{Z}[\{1\}]$ component of the selfintersection numbers are equal.*

Sketch of proof: To see the invariance of $\lambda(A, B)$ consider a homotopy $A \times I \to M$. Make the homotopy transverse to the other surface, then the inverse image is a collection of arcs and circles in $A \times I$ whose endpoints are intersection points of $A \times \{0, 1\}$ with B. An arc with both ends on one end of $A \times I$ gives intersection points which cancel algebraically.

An arc with ends on opposite ends gives the same contribution to each intersection number. Since all intersection points occur as endpoints of arcs, the intersection numbers must be equal.

For selfintersections note that a regular homotopy is a sequence of isotopies, Whitney moves, and finger moves. Isotopies do not change intersections. Finger moves introduce two intersection points which cancel algebraically, so do not change the selfintersection number. Conversely Whitney moves remove two such intersection points.

According to 1.6 essentially the only other things encountered in a general homotopy are cusps and their inverses. These remove or introduce a single selfintersection point with trivial associated loop, so can change the number only by elements of $\mathbf{Z}[\{1\}]$.

For the converse suppose A and A' are homotopic rel boundary and have the same framed boundary. Passing through a cusp changes the Euler number of the normal bundle ($A = S^2$) or the rotation number at the boundary relative to a fixed framing ($A = D^2$), and these show up in changes in the $\mathbf{Z}[\{1\}]$ component of the selfintersection number. If these components are equal the number of cusps, counted with signs, must be zero. Since two cusps with opposite sign can be cancelled to leave a regular homotopy (see the picture below), all cusps can be cancelled and A, A' are regularly homotopic.

 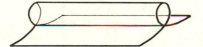

1.8 Sums

Suppose A, B, and C are immersed surfaces, and c is an embedded arc in C from a point $A \cap C$ to a point $B \cap C$. Then the *sum* $A \# B$ of A and B along c, is obtained by deleting disks in A and B about the intersection points, and adding the linking annulus of the arc c.

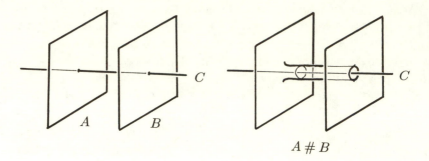

$$A \# B$$

(Here we have omitted the extension of C to past and future, using the product convention.)

If A and B are distinct, then $A \# B$ is also framed immersed. The orientation of the framing on one of them may have to be changed, however, in order to extend to a framing of the sum. If $A = B$, the result is denoted by $A\#$, and is also referred to as doing a *0-surgery* on A. If the intersection points have opposite sign (in the sense of 1.4) then the framing of A extends to a framing of $A\#$.

Exercise Suppose $A \to M$ is a framed immersed sphere with $\mu(A) = 0$. Show there is a framed embedded surface $A' \in M$ representing the same homology class, with $\pi_1 A' \to \pi_1 M$ trivial. (Note A and any $A\#$ represent the same homology class, since they differ only in a neighborhood of a 1-complex.) □

Intersection numbers behave nicely with respect to sums: suppose A is an immersed sphere, B, C disks or spheres. Suppose there is an embedded arc from A to B which represents $g \in \pi_1 M$. Form the sum along this path $A \# B$ in such a way that the orientations of the pieces agree with an orientation of the result. Obtain a path to the basepoint from the result by using the path originally chosen for B. Then $\lambda(A \# B, C) = \lambda(B, C) + g\lambda(A, C)$.

1.9 Transverse spheres

A *transverse sphere* for a connected surface A is a framed immersed sphere whose image intersects A in exactly one point. If A is not connected then *transverse spheres* for A are transverse spheres for each component disjoint from the other components. We will use a superscript t to denote transverse spheres: A^t.

Note that if there is one transverse sphere for A then we can get arbitrarily many by taking parallel copies.

Sums with transverse spheres will be used as a way to remove intersections. As a typical application, suppose that $A \cup B \to M$ is a framed
immersion, and A has a transverse sphere A^t. Then adding parallel
copies of A^t to B at each intersection point $A \cap B$, along arcs in A,
yields a surface B' disjoint from A.

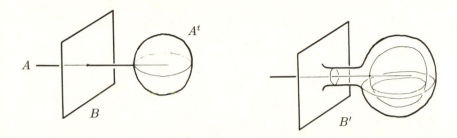

Exercise Suppose A is a framed immersed disk in M, with an *embedded* transverse sphere. Show that there is an embedded disk in M with
the same framed boundary. □

As illustrated by the exercise, embedded transverse spheres are very
useful. They are also very rare. There is a "stabilized" version of 4-
dimensional topology which is much easier because one gets embedded
transverse spheres for free. Define $M \# (S^2 \times S^2)$ by deleting small
open balls from M and $S^2 \times S^2$ and identifying the resulting 3-sphere
boundaries. Note that there are two framed embedded S^2 in $S^2 \times S^2$
which intersect in one point. If A' is obtained by adding one of these
spheres to a surface A in M (along an arc) then A' has an embedded
transverse sphere. Combining this observation with the exercise yields:
if A is an immersion of a union of disks in M then for some k there are
embedded disks in $M \# k(S^2 \times S^2)$ with the same framed boundaries.

Immersed transverse spheres are not much good for removing selfintersections, but can be used to make two things disjoint. As with the
immersed Whitney move this disjointness comes at the expense of increasing the number of selfintersections. The use of transverse spheres
is slightly more violent than immersed Whitney moves in that the latter
changes things by regular homotopy, whereas the former often does not.

An *algebraically transverse sphere* for a connected immersed surface
A is a framed immersed sphere B with algebraic intersection $\lambda(A, B) =
1$. Generally if A has components A_i then "algebraically transverse
spheres" for A is a framed immersion of a union of spheres $\cup B_i \to M$,
so that $\lambda(A_i, B_j) = 1$ if $i = j$, and $= 0$ if $i \neq j$. Existence of algebraically transverse spheres is a purely algebraic-topological condition,

and occurs frequently. We note that such A and B can be changed by regular homotopy rel boundary—using the immersed Whitney move—to be transverse in the geometric sense.

The final topic in this section is a characterization of surfaces which have transverse spheres. First, we say that an immersion $A \to M$ is π_1-*negligible* provided the complement of the image has the same fundamental group; $\pi_1(M - A) \to \pi_1(M)$ is an isomorphism. We will see this condition is equivalent to the existence of a possibly unframed immersed sphere intersecting A in a single point.

To detect the framing, consider the second Stiefel-Whitney number as a homomorphism $\omega_2 \colon \pi_2 \to \mathbf{Z}/2$. An immersion is said to be *dual to ω_2 on π_2* if the mod 2 intersection numbers with A define the same homomorphism. Conversely, A is *not* dual to ω_2 if these homomorphism are different.

Proposition. *A connected framed immersed surface in a 4-manifold has a transverse sphere if and only if it is π_1-negligible, and is not dual to ω_2 on π_2.*

Proof: Let M_0 denote the complement of the interior of a regular neighborhood of A. The boundary of a normal disk to A defines a framed embedding of S^1 in ∂M_0. This circle is nullhomotopic in M—it bounds the normal disk—so if A is π_1-negligible it is also nullhomotopic in M_0. Apply the immersion lemma to replace the nullhomotopy by an immersion of a disk, which on the boundary differs from the given immersion by rotations. If the difference is an even number of rotations, twist in the interior of the disk to match up the framings. The resulting immersion fits together with the normal disk to give a transverse sphere for A.

If the immersions differ by an odd number of twists (which happens when $\omega_2 \neq 0$, according to 1.3), the immersions fit together to give a immersion with nontrivial normal bundle. An unframed transverse sphere implies that A is π_1-negligible, since sums with it move disks off A. Note however that sums with an unframed sphere destroys framings. This but is why a framing is included in the definition of transverse spheres.

If the immersions differ by an odd number of twists, then since A is not dual to ω_2 there is an element $f \in \pi_2 M$ with $\omega_2(f) = 1$ but $\lambda(f, A) = 0$ mod 2. Represent f by an immersion in general position with respect to A, and use sums with the unframed transverse sphere to obtain f' disjoint from A. Since the intersection number is trivial mod 2, an even number of copies of the sphere are added to f, so $\omega_2(f') = 1$. Adding f' to the unframed transverse sphere gives a sphere intersecting A in one point, but with $\omega_2 = 0$. According to the first part of the argument this sphere is homotopic to a framed immersed transverse sphere. ∎

1.10 Tracking intersections

Some of our constructions involve large numbers of surfaces, with complicated intersections. The most common error is to think that two surfaces are disjoint, when in fact they intersect. To avoid this we use two conventions: the description of new surfaces as "constructed from" parallel copies of old ones, and the use of "scorecards".

In the construction descriptions we take advantage of the fact that surfaces are generally disjoint from 1-complexes. Therefore modifications which take place in a neighborhood of 1-complex will not change intersections (with certain obvious exceptions).

To illustrate this, suppose we have A, B, and an immersed Whitney disk W for a pair of intersections $A \cap B$. Let A' be the surface obtained by an immersed Whitney move of A across W. A' is described in 1.4 as constructed from A and two parallel copies of W (by some cutting and pasting in a neighborhood of the Whitney circle). The plumbings used to define framed immersions do not allow for intersections of anything else with a Whitney circle. Therefore manipulations near the Whitney circle effect only the two intersection points which lie on the circle. All other intersections of A' come from the component surfaces.

Note that in sums and finger moves the entire modification takes place in a neighborhood of an arc. Therefore these cause no change in intersections other than the obvious ones.

A *scorecard* is a table in which the intersection data is recorded. For example suppose we have disjoint immersed disks A, B, and a framed immersed sphere C which is algebraically transverse to A. This data is summarized by:

	A	B	C
A	?	ϕ	alg. 1
B	ϕ	?	?
C	alg. 1	?	?

Here a question mark indicates no restriction are known. We indicate a sequence of moves which produces A', B', C' with A' and B' still disjoint and with C' a (geometrically) transverse sphere for A'. Choose immersed Whitney disks W_i for all but one of the intersections between A and C (using 1.6). We then have intersections

	A	B	C	W_i
A	?	ϕ	alg. 1	?
B		?	?	?
C			?	?
W_i				?

(Duplicated entries have been omitted.) Obtain C' by using finger moves to pushing off of int(W_i). Obtain A' by immersed Whitney moves on A, using the Whitney disks W_i. Since A' is constructed from A and parallels of the W_i, we have scorecard

	A'	B	C'
A'	?	?	1 pt.
B		?	?
C'			?

C' is now a transverse sphere for A', (and usually would be renamed $(A')^t$). Add copies of C' to B along arcs in A' to remove intersections with A'. This yields a disk B' as required;

	A'	B'	C'
A'	?	ϕ	1 pt.
B'		?	?
C'			?

1.11 References

A general reference for elementary differential topology (eg. lemma 1.2) is Guillemin and Pollack [1]. For PL topology see Rourke and Sanderson [1]. Wall [1, p.45] is a reference for intersection numbers in nonsimply connected manifolds.

Twisting, to correct the framing of an immersion, was introduced in Freedman and Kirby [1]. The picture of it used here comes from Quinn [1]. The appearence of twisting, in the form of cusp homotopies, follows from Whitney [2].

Whitney disks, which are generically embedded in dimensions greater than 4, were introduced by H. Whitney [1] in 1944 to show n-manifolds can be embedded in $\mathbf{R}^{2n}$. Their role as one of the workhorses of higher dimensional topology has been clear since their use by Smale [1] in 1959 to prove the generalized Poincare conjecture. The role of immersed Whitney disks in dimension 4 has emerged more slowly. Early occurrences are in Casson [1] and Kobayashi [1]. Embedded Whitney disks in a stable setting (mentioned after exercise 1.9) were developed by Quinn [1] and by T. Lawson.

The use of finger moves to make things disjoint is often called "piping" in higher dimensions, and was described by Penrose, Whitehead, and Zeeman [1]. In dimension 4 it has been long known to imply that homotopy implies isotopy for 1-complexes (eg. Andrews and Curtis [1]). More sophisticated use of the technique was made by Casson [1], see the appendex to Freedman [1].

Sums with a transverse sphere, as a method to remove intersections, was pioneered by Norman [1], in 1969. This idea has figured in most of the delicate work done since then, eg. Casson [1]. Many of the terms and basic properties used here were set out in Quinn [1]. "Self-sums", or 1-surgeries, were used in Freedman and Quinn [1]. This technique has had little use until recently because the utility of surfaces of higher genus was not appreciated.

CHAPTER 2

Capped gropes

Capped gropes are 2-complexes with the same regular neighborhood as a fixed surface, usually a disk. Intersections which are "more subtle" than the standard plumbing points can be defined in terms of this more complicated description of the neighborhood.

The first three sections describe capped surfaces, and this construction is iterated in 2.4 to define capped gropes. The key property of gropes is that once a certain height has been obtained they can "grow." This is proved in 2.7, and then exploited in the remainder of the chapter. It is used to disengage the image of gropes from the fundamental group of the manifold, and in a controlled setting to disengage images from each other.

2.1 Capped surfaces

Consider a neighborhood of a framed disk. Other surfaces will usually intersect it (according to the plumbing construction and lemma 1.3) in normal 2-disks. A more subtle sort of intersection occurs when two surfaces enter and leave the regular neighborhood in an inessential way, but "link" each other.

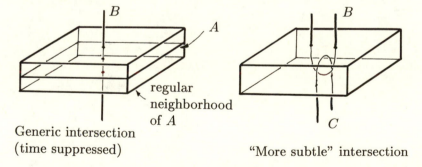

Generic intersection
(time suppressed)

"More subtle" intersection

We formalize the "more subtle" intersection this way: Add A to itself to obtain a surface with one handle, disjoint from B and C. (Push A below B, and use 1.9 to remove $A \cap C$.) There are two canonical basis curves on the surface which bound disks in the neighborhood, one intersecting

B and one intersecting C. These curves intersect each other in exactly one point.

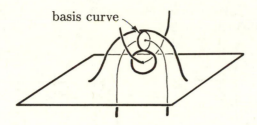

basis curve

Now suppose S is a surface. An S-*like capped surface* is a surface in $S \times \mathbf{R}^2$ obtained by replacing disks in S with copies of the picture. This gives a surface—the *body*—obtained by doing some number of self-additions on $S \times \{0\}$, together with canonical disks spanning the canonical basis curves. The disks are called the *caps* of the capped surface. Caps occur in pairs with a point in common; the *dual* of a cap is the other member of this pair. Caps have boundary on the body surface, in the sense of 1.3.

We will refer to the surface S as the "starting surface" to avoid confusion between S and the body. For the most part S will be a disk, or a union of disks.

There is another description of this construction in terms of handlebodies. Consider the product structure on $S \times I$ as a handlebody structure with no handles. Introduce some cancelling pairs of 1- and 2-handles. Then the "body" surface is the level surface between the layers of handles, and the caps are the cores of the 2-handles and the dual cores of the 1-handles. The 4-dimensional model is obtained by product with I.

These definitions specify a regular neighborhood of a capped surface, as well as the object itself. Note that the regular neighborhood is isomorphic with $S \times D^2$. Also note that capped surfaces have models in $S \times \mathbf{R}$, and the 4-dimensional versions are obtained by a further product with $\mathbf{R}$. This fact will be very useful to us, since it means we can get disjoint parallel copies in 4-space by product with different points in the last coordinate.

An immersion of a capped surface in a 4-manifold is defined as in 1.2: introduce plumbings into the 4-dimensional model, and fix an isomorphism of the result with an open set in the manifold. Note that—as with surfaces—this definition of immersion requires that the boundary of the body be taken to the boundary of the manifold.

Example 1. Suppose A, B are immersed surfaces with an algebraically

cancelling pair of intersection points, with an immersed Whitney disk
W. Add A to itself along the boundary arc of W on B. Caps for the
resulting surface are obtained from W and a small disk cutting across
the sum tube. This gives an A-like capped surface whose body has two
fewer intersection points with B.

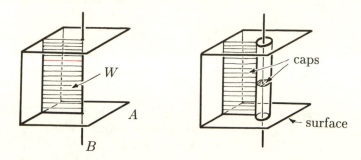

Exercise   An immersion of a disk-like capped surface gives a framed
immersion of the body surface which is trivial on the fundamental group
(the caps give nullhomotopies of the generators). Show that conversely
a framed immersion of a compact orientable surface with one boundary
component, trivial on fundamental group, extends to an immersion of a
disk-like capped surface. (See 1.4, and corollary 2 of 1.3.) □

Example 2.   A Whitney disk has a transverse sphere-like capped surface; The linking torus of one of the intersection points is an embedded
torus intersecting only the Whitney disk, and in only one point, and
normal disks for the two intersecting surfaces provide caps. The intersections between the caps and the surfaces are usually removed using
transverse spheres (see the proofs of 5.2, 5.3).

   We explicitly describe this capped surface, continuing the description
of the torus in exercise 1.1. If a neighborhood of the intersection point
is parameterized as $\mathbf{R}^2 \times \mathbf{R}^2$, with the Whitney disk the quadrant $\mathbf{R}_+ \times
\{0\} \times \mathbf{R}_+ \times \{0\}$, then the torus is $S^1 \times S^1$ and the caps are $D^2 \times \{p\}$ and
$\{p\} \times D^2$. Here $p \in S^1$ is some point with nonzero second coordinate.

2.2 Proper immersions

   A *proper* immersion of a capped surface is an immersion in which the
body has no selfintersections or intersections with the caps. A proper
immersion can be thought of as an immersion of the starting surface in
which all the selfintersections are "more subtle," in the sense discussed
in 2.1.

If an immersion of a disk with boundary in ∂M has algebraic selfintersection 0 then we can use the method of the example in 2.1 to get
an immersed capped surface with body embedded. This is not proper
since the caps still intersect the body. Typically these intersections are
removed with a transverse sphere.

Exercise (1) Suppose A is a framed immersed disk with $\mu(A) = 0$,
and an immersed transverse sphere A^t. Show there is a properly immersed disk-like capped surface with the same framed boundary, and
a transverse sphere for the body. (2) Suppose the hypotheses of the
main embedding theorem: an immersed disk A, a sphere B algebraically
transverse to A, and $\mu(B) = 0$. Show that there is a properly immersed
capped surface with the specified framed boundary, and a transverse
sphere for the body which is disjoint from the caps (start with the observation at the end of 1.8). □

The regular neighborhood of the image of a capped surface should be
thought of as a pair: the neighborhood itself, which is a 4-manifold, and
the regular neighborhood of the boundary of the surface, in the boundary. The neighborhood of the boundary has a canonical description as
$S^1 \times D^2$. The structure of the pair is a bit complicated. However the
structure of the neighborhood by itself (forgetting the position of the
$S^1 \times D^2$) is quite simple.

Suppose (as in 1.2) the starting surface has boundary in each component, so is a regular neighborhood of a graph. At each intersection point
between caps add an arc in each sheet joining the intersection point to
the graph. Then the regular neighborhood of the image is also a regular
neighborhood of this 1-complex. To see this, note first that the image
collapses to the image of the caps, the graph in S and embedded arcs
joining them. Then apply the corresponding fact for the caps (1.2).

As in 1.2 this implies that the fundamental group of the image is free,
and is generated by loops passing through at most one intersection point
if the starting surface is connected.

2.3 Contraction

This essentially recovers S from an S-like capped surface, so provides
a converse to the construction in the example of 2.1.

Cut a capped surface open along the attaching curves of a dual pair
of caps. Fill in with two parallel copies of each cap, and a small central
square.

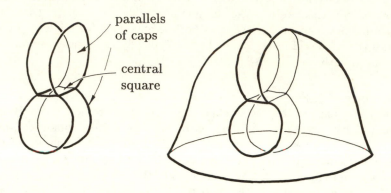

Doing this to each pair of caps yields a surface, called the *contraction* of the capped surface. The contraction is defined in a small neighborhood of the model capped surface, so an immersion of a capped surface defines—by restriction—an immersion of the contraction.

Notice in the first example in 2.1 the starting surface is exactly recovered by cutting and adding copies of only half the caps. In the model there is an isotopy from this "half contraction" to the contraction as defined above. If an immersion is given then this isotopy defines an regular homotopy from the image of the "half contraction" to the image of the contraction. Therefore contraction applied to the construction of 2.1 yields an immersion of the surface regularly homotopic to the original. This is the sense in which contraction is inverse to this construction.

It also follows from this discussion that the regular homotopy class of the contraction is independent of the caps: if two immersions of the caps are given, form a "hybrid" with half the caps from one set, the other half from the other. Since the regular homotopy class of the contraction only depends on half the caps, each contraction is regularly homotopic to the contraction of the hybrid.

Next suppose an immersion of a capped surface is given. Any surface which intersects a cap will have two corresponding intersections with the contraction. Further these two points have a Whitney disk. Choose an arc on the cap from the original intersection point to the point of intersection with the dual cap. This defines a ribbon between the parallel copies of the cap, from the new pair of intersections to the central square. This ribbon is the Whitney disk.

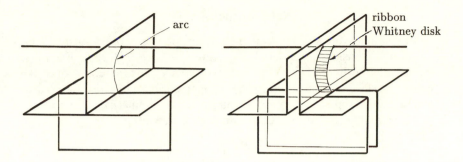

Since this disk is defined in a neighborhood of an embedded arc, it will be embedded disjointly from other surfaces in the manifold. If there are many intersections with a single cap we can choose an arc for each one, all disjoint except for the endpoint. These yield disjoint Whitney disks. However arcs on dual caps give Whitney disks which intersect; the Whitney circles intersect as they cross the central square.

Suppose $A \to M$ is an immersed capped surface, and $B \to M$ is a surface in general position with respect to A and intersecting only the caps. The phrase *push B off the contraction of A* means: contract A, construct ribbon Whitney disks for the intersections of the contraction with B, disjoint except for the necessary points on the central square, and push B across these Whitney disks. This yields a surface B' disjoint from the contraction. The only new intersections of B' with anything are selfintersections from the intersections of the Whitney circles; when B is pushed across a Whitney disk from an arc on one cap, it passes through Whitney disks from the dual cap. Pushing across these then give intersections. We depict these intersections schematically as follows:

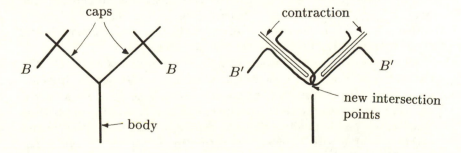

2.4 Capped gropes

Models for gropes are obtained by iterating the capped surface construction. Consider a model S-like capped surface in $S \times I$. Each cap has a neighborhood $D^2 \times I$, which we replace with a model disk-like capped surface. These second stage capped surfaces likewise have caps, which can be replaced by capped surfaces. We define *capped gropes* to be objects obtained after a finite number of such steps.

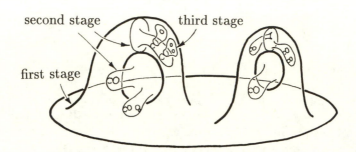

There is some terminology similar to that used for capped surfaces. The caps of the last step in the construction are the *caps* of the grope, the union of all the surfaces constitutes the *body*. The body is divided into *stages*: the original surface is the first stage, the surfaces replacing the first stage caps are the second stage, etc. We say that a capped grope has *height at least* n if the replacement has been done at least $n - 1$ times; the body has at least n stages. A capped surface, for example, is a capped grope of height 1. Finally note that if the caps of a capped grope are all replaced by disk-like capped gropes, the result is another capped grope. The new one is called an *extension* of the original.

Immersions of capped gropes are defined as always, by introducing plumbings into the model. A *proper* immersion is one in which plumbings are allowed only among the caps. This restriction implies that the body is embedded, disjointly from the interiors of the caps. As with surfaces it is also a consequence of our definition of "immersion" that the boundary of the first stage goes to the boundary of M.

We caution against a confusion which may occur in rare cases. Usually the body and caps can be identified by topological properties; caps are disks with nothing attached to them. This is the case when in the model at each stage the caps are replaced by nontrivial capped surfaces. However a disk is trivially a disk-like capped surface with no caps, and if this is used as a replacement we get a disk as part of the body. For example we can think of a k-stage capped grope as a $(k+1)$-stage grope with trivial top stage. This makes a difference in what is allowed in a proper immersion: as a $(k+1)$-stage grope the previous caps are now part of the body, so a proper immersion would have to be an embedding.

Contractions are defined for capped gropes. By contracting the topmost capped surfaces in a capped grope of height at least n, we get a capped grope of height at least $n-1$. Contracting stage by stage eventually yields the starting surface. An immersion of a capped grope defines immersions of all its contractions. Note particularly that if the original immersion is proper, the contractions are also properly immersed.

One benefit of having many stages in a grope is that we can get a little extra disjointness when pushing off contractions. Suppose for example that A is a properly immersed capped grope with at least n stages, and suppose a surface B intersects A only in the caps. Suppose for each of these intersections there is a designation of one of the upper n stages of the grope. Contract the top stage and push off all intersections whose designated stage is the top one. Then contract the next stage, and push off all intersections which have that as their designated stage. Continue contracting and pushing off a total of n times. At the end we have a surface B' disjoint from the n-fold contraction. The new intersections in B' come from intersections of B with A which were pushed off the contraction at the same stage. Pieces pushed off at different stages are disjoint. This operation will be used in 2.9.

Exercise Suppose A is a properly immersed capped grope of height n. In a neighborhood of the image construct arbitrarily many disjoint properly immersed capped gropes of height $n-1$ with bodies parallel to the lower $n-1$ stages of A. (Push A off the contraction of a parallel. These are called *disjoint near parallels*, of the contraction, since the bodies are parallel to that of the contraction.) ◻

2.5 Pushing intersections down

Suppose a surface B intersects a capped grope, say in stage n. We can do a finger move to push B off stage n and through stage $n-1$. B will intersect stage $n-1$ in two points. Now do two finger moves to push these off through stage $n-2$, and get 4 intersection points.

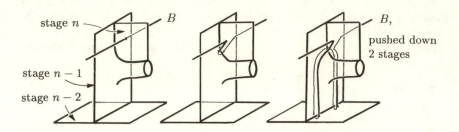

Pushing down k stages produces a cascade of 2^k little fingers. Notice that there are disjoint paths to push down along, because the complement of the attaching curves on each surface is connected. Pushing down is contrary to the idea—which we have tried to encourage by our use of the term "proper"—that we should only allow intersections with the caps. It is usually done only when the new intersections can be removed quickly, for example with a transverse sphere for a lower stage.

Exercise   Show that under the hypotheses of the main theorem there is a properly immersed disk-like 3-stage capped grope with the specified framed boundary, and with a transverse sphere for the bottom grope stage disjoint from the rest of the capped grope. (Continue from exercise (2) in 2.2. Review the proof of that result for hints, if necessary.) ☐

2.6 Transverse gropes

Suppose $A \to M$ is an immersion of a connected capped grope. A *transverse grope* for A is a properly immersed sphere-like capped grope A^t so that there is a single point of intersection of the bottom stage of A with the bottom stage of A^t, and all other intersections are between caps. If A is not connected, a transverse grope is a properly immersed (union of spheres)-like capped grope, with a component transverse to each component of A, and (other than the transverse points) only intersections between caps.

Notice that total contraction of transverse gropes gives transverse spheres for the bottom stage surfaces, if the height of A is at least one. These spheres may intersect the caps of A. If A has height 0 (ie. is an immersed surface) then contractions yields algebraically transverse spheres.

We construct transverse gropes inside models of capped gropes. Let S be a surface, and consider a model S-like capped grope of height n in $N \times I \times [-2, 2]$. Collect the second stage surfaces into two "sides", denoted E_+ and E_-, so that the surfaces on each side are disjoint from one another. For example in the 3-dimensional model for a capped grope we could choose the $-$ surfaces to be the ones on the 0 side of the first stage surface, and the $+$ ones to be on the 1 side. The terminology "side" comes from this picture. Let F be the union of E_- and all higher stages attached on it. Then F is a (union of disks)-like capped grope of height at least $n-1$. $\partial F \times [-1, 1]$ is a collection of annuli which intersect $F \times \{0\}$ in its boundary. Let G be an annulus obtained by perturbing the interior of $\partial F \times [-1, 1]$ to be disjoint from F. Finally define E_+^t to be $F \times \{1\} \cup F \times \{-1\} \cup G$.

In a neighborhood of the intersection point $E_+ \cap E_-$ this looks like

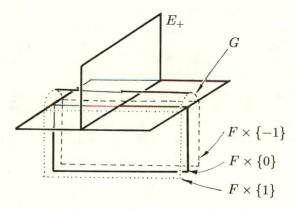

E_+^t is a (union of sphere)-like grope which intersects E_+ in exactly one point in each component. In fact the intersection point with E_+ is the only intersection with the whole grope. When the model is immersed, however, cap intersections lead to intersections between the caps of E_+^t and the caps of the grope.

There is a transverse sphere-like grope E_-^t for the other side defined the same way. Unfortunately corresponding components of E_+^t and E_-^t intersect in two points in their bottom stages. This will usually prevent us from using them both at once.

Exercise   Draw both E_+^t and E_-^t in a neighborhood of the intersection point, as above. Find the two points of intersection. ☐

2.7 Raising grope height

The next proposition is the first of several "replication" results. These state roughly that once the data has been refined to a certain point, it can automatically be refined further.

Proposition. *Suppose there is a proper immersion of a capped grope of height at least 3 in M. Then for any n there is an extension of the body to a proper immersion of an n-stage capped grope.*

We recall that an extension is obtained by replacing the caps by capped gropes (2.4). Note the new grope has the same framed boundary as the original, and up to regular homotopy the same starting surface, because these are both determined by the first stage.

Proof: Divide the second stage surfaces into $+$ and $-$ "sides" as in 2.6. We show how to raise the height of the $+$ side. Proceeding in steps alternately raising the height of the $+$ and $-$ sides will then prove the proposition.

Form transverse gropes E_+^t for the second stage $+$ side, as in 2.6. These are constructed from parallel copies of the $-$ side, so have height at least 2. Contract the top stage of the transverse gropes, and push the caps of the original grope off the contraction. The new transverse gropes are disjoint from the caps.

Now at every intersection point of a $+$ cap with any other cap, push the other cap down to the second stage, and add parallel copies of the transverse gropes. This converts the caps into capped gropes. Since the transverse gropes are disjoint from the caps, these capped gropes are properly immersed disjointly from the body of the original grope. Therefore they give a proper extension with greater height on the $+$ side. ∎

Exercise Weaken the hypothesis in the proposition to: the grope has height at least 1 on the $+$ side, at least 2 on the $-$ side. (For this form E_+^t, push $+$, $-$ cap intersections down the $+$ side, and add copies of E_+^t. This raises height on the $-$ side, but the immersion is not proper since the $-$ caps may intersect the transverse grope caps. Contract the new $-$ grope stages and push off the $+$ caps. This gives new caps with the $+$ side disjoint from the $-$ side. Repeat the construction, and observe that with this extra disjointness the immersion is proper. Use scorecards.) □

2.8 Speed, complexity, and size of height raising

We will need estimates on how various parameters of gropes change in the proof in 2.7.

Speed. Begin with a grope of height 3, and proceed in steps; in step n raise the height of the $(-1)^n$ side. Let a_n denote the height of this side after the step. The height after a step is the original height, plus the height of the contracted transverse gropes. The transverse gropes are constructed from all but the bottom stage of the opposite side, so we obtain the recurrence relation $a_n = a_{n-1} + a_{n-2} - 2$. With initial grope height 3 we have initial conditions $a_{-1} = 3, a_0 = 3$. These conditions identify the heights as being essentially the Fibonacci numbers. Explicitly we can work out

$$a_n = 2 + \frac{\sqrt{5}+3}{2\sqrt{5}}\left(\frac{1+\sqrt{5}}{2}\right)^n + \frac{\sqrt{5}-3}{2\sqrt{5}}\left(\frac{1-\sqrt{5}}{2}\right)^n,$$

but we will only use the easy estimate $2^{n/2} \leq a_n \leq 2^n$. Fibonacci numbers are widely observed in the growth of plants, so their occurrence here provides amusing evidence for the "organic" nature of gropes.

Complexity. The complexity of an immersed grope is measured in terms of words in the fundamental group of the image. For this estimate we assume the grope is disk-like, though more general versions are easily obtained.

Let G_i denote a regular neighborhood of the image of the capped grope produced in step i of the construction. Then $\pi_1 G_i$ is a free group with canonical generators given by loops passing through exactly one intersection point. The estimate is that under the homomorphism $\pi_1 G_{i+1} \to \pi_1 G_i$ generators go to words of length at most 5 in generators and their inverses. Thus after n steps $\pi_1 G_n \to \pi_1 G_0$ takes generators to words of length $\leq 5^n$ in the original generators and their inverses.

The transverse gropes are parallel to one side of the grope, so generators in these go to original generators. Contracting the top stage does not change the set of elements which occur as generators, although it increases the number of intersections. Pushing the original caps off the contraction introduces new intersections among the caps. Loops for these intersections pass through 2 old intersection points, so these generators go to words of length 2. (See the picture in the next section.)

In the final result the longest words occur as loops passing through intersections in the new + side caps. Such a loop passes through an ++ or +− intersection point, through a pushdown finger, then out through the transverse grope to the cap. It passes through the new intersection point (between caps of transverse gropes) and finally back to the basepoint along a similar path. The loop passes through a total of three intersection points. The first and last may contribute length 2 in the original generators, the middle contributes 1. The maximum length in terms of the original generators is therefore 5, as claimed.

Size. Sizes are measured in terms of diameters of images in a metric space. Suppose X is metric, $f : M \to X$ is continuous, and $\delta > 0$. Suppose that the image in X of each component of the grope has diameter less than δ. The estimate is that each step in the height raising argument may increase diameters by a factor of 7. Thus after n steps in the proof, component images have diameter less than $7^n\delta$.

This estimate is used in the controlled embedding theorem. In contrast to the complexity estimate it is most useful when the grope has lots of components.

The transverse gropes have the same initial diameter as the pieces from which they are constructed. Contraction does not increase diameter. Pushing off the contraction involves pushing across ribbon Whitney disks which may have diameter δ, and so may increase the diameters of the caps to 3δ. (Note changes get doubled: pushing across a Whitney disk may move points as far as δ from the original image. Applying the triangle inequality shows that points moved by different Whitney disks are within 3δ.) When pushing down and adding copies of the transverse gropes., the pushdown fingers may have diameter δ, and the transverse gropes have diameter δ. This may move points as far as 2δ, so final diameter is bounded by 7δ, as claimed. ∎

Exercise (1) Estimate the height, complexity, and size growth rates for the more complicated version of substep 1 used in the exercise in 2.7. (2) Construct disjoint near parallels for properly immersed capped gropes of height at least 2, without loss of height. (See the exercise in 2.4). ☐

2.9 Reducing the fundamental group

The fundamental group restriction in the main embedding theorem comes from a step which requires finding an immersed capped grope with the property that loops in the image are contractible in the ambient manifold. The following is the best which can be done at the present time in this direction.

Proposition. *Suppose* $\alpha : M \to \Gamma$ *is a homomorphism,* Γ *is a poly-(finite or cyclic) group, and* M *contains a properly immersed disk-like capped grope of height at least 3. Then there is a properly immersed capped grope with the same body, so that the fundamental group of the image is taken to* $\{1\}$ *by* α.

Proof: A poly-(finite or cyclic) group is one with a finite ascending sequence of subgroups, each normal in the next, so that successive quotients are either finite or infinite cyclic. It is sufficient to prove the result when Γ itself is finite or infinite cyclic. The general case will follow by induction on the number of subgroups in the sequence, applying the special case to the quotients.

Γ **finite.** Let n be the order of Γ, and apply Proposition 2.7 to raise the height of the grope by n. Choose a correspondence between elements of Γ and the new grope stages. We contract the new grope stages, and push intersections among the caps off in a careful way.

At each intersection point choose a first and second sheet. This defines an element in the fundamental group of the image: the loop going through only this intersection point, passing from the first to the second sheet. Then apply the correspondence of Γ with the new grope stages. This gives for each cap intersection a designation of one of the top n grope stages. Now contract the top stage, and push off (first sheet off second) all intersections whose designated stage is the top one.

When the caps are doubled in the contraction, intersection points are multiplied by four. Replicate the data for these new intersections by letting the first sheet be the parallel of the old first sheet and the designated stage of the grope be that of the old intersection point.

Now repeat the construction: again contract the current top stage and push off all intersections which have this as their designated stage. Repeat this process n times. At the end the new grope stages have all been contracted, so we have new caps for the original grope body. Since we have "pushed off" all cap intersections during the construction, all intersections among these new caps come from the pushing off process. These arise when (decendents of) old intersections are pushed off dual caps in a single stage of the contraction (2.3). If these intersections have fundamental group elements g, h respectively, then the new intersection points have group element gh^{-1}.

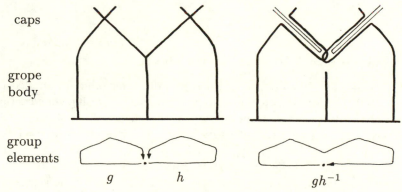

We note that the doubling of intersections in contractions does not change the associated group element. Therefore if two intersections are pushed off the same stage of the contraction, their loops must have the same image in Γ, say i. Since α is a homomorphism, the new intersection points have loops with image $ii^{-1} = 1$. Since these loops are generators for the fundamental group of the image of the grope with the new caps, the entire fundamental group is taken to $\{1\}$ by α.

Γ **infinite cyclic.** We can no longer afford to have a separate grope stage for each group element which may occur, so we describe a more elaborate strategy. We use the additive notation for elements of the infinite cyclic group $\mathbf{Z}$.

Suppose that the the the integer N is a bound for the absolute values of the image in $\mathbf{Z}$ of the set of generators for the fundamental group of the grope image. Contract the top stage, and push off all intersection points with loop image between $N/2$ and N. Contract the next stage and push off all between $-N$ and $-N/2$. The generators in this new immersion have image bounded by $N/2$: the old ones by choice and the new ones since they are differences in one of the two cases.

Iterating this construction shows that $2i$ stages can be used to get generators whose images in are bounded by $(1/2)^i N$. In particular if i is large enough so that $(1/2)^i N < 1$, then the image is trivial.

Now recall the estimates given in 2.8. Suppose we start with a grope of height 3 whose generators have images bounded by B. After $2k$ steps we have words of length 5^{2k} in the original generators, so the image is bounded by $5^{2k} B$. The grope height is at least 2^k. The strategy above can therefore be used to give trivial image after $2k$ steps if

$$\left(\frac{1}{2}\right)^{2^k} \left(5^{2k}\right) B < 1.$$

But since

$$\lim_{k \to \infty} \left(\frac{1}{2}\right)^{2^k} \left(5^{2k}\right) = 0,$$

there exists some k for which this will work. ∎

Exercise (1) Suppose the original grope has image in $\mathbf{Z}$ bounded by $B = 100$. Determine how many steps in the height raising construction are necessary to get a grope for which the image-killing strategy will succeed. (2) See that the strategy fails for Γ a free group on r letters, $r > 2$. (The number of group elements represented by words of length k is nearly $(2r)^k$, and the only strategy we see for getting rid of them is to use a separate stage for each group element.) (3) Verify the remark in the introduction that we can kill images in the following class of groups: the class containing finite and cyclic groups, and closed under direct limits, subgroups, quotients, group extensions, and passage to a larger group in which the original has finite index. □

2.10 Controlled separation

This material is required for the controlled embedding theorem, but not for embedding a single disk.

Suppose X is a metric space, and $f : M \to X$ is a map from a 4-manifold to X. We want to construct gropes, etc. in M and control the size of the images in X. There are two main ways to do this. The first is to show that the construction requires certain fixed number of steps, each of which has a predictable effect on the size. The second is to separate the data for the construction into small disjoint units. After this arbitrarily many steps can be used, and the results will remain inside regular neighborhoods of these units. The object of this section is to use the first method to separate data into units, so that the second method can be applied.

Lemma. *Suppose X is a locally compact metric space, and $\epsilon: X \to (0, \infty)$ is given. Then there is $\delta: X \to (0, \infty)$ such that if $M^4 \to X$ is continuous, and $A \to M$ is a properly immersed (union of disk)-like capped grope of height at least 3 such that the images of the components of the grope in X have diameter less than δ, then there are new caps for the bodies such that distinct components of the grope are disjoint in M and the images in X have diameter less than ϵ.*

When ϵ is a function what we mean by "diameter less than ϵ" is that the diameter is less than $\epsilon(x)$ for all x in the set. Although control functions are necessary when X is not compact, they cause a lot of essentially irrelevant technical trouble. We will therefore begin with X compact and ϵ constant.

Proof, compact case: Suppose the components of the grope are divided into two collections. Then we could separate these collections as follows: raise height by 1. Contract all the new stages in the first collection, and push off caps of the second collection. Then contract the second collection.

If the images in X start out smaller than δ, then raising the height gives diameter $7^2\delta$ (by the size estimate in 2.8; 2 steps are required, one on the $+$ side one on the $-$). Contracting does not increase diameter, but pushing off may triple diameters in the second collection. The entire construction therefore yields diameter less than $3(7^2)\delta$.

More generally we could separate k collections by contracting the first and pushing the others off, contracting the second and pushing the later ones off, etc. The last collection would be pushed off $k - 1$ times, so would have diameter less than $3^{k-1}(7^2)\delta < 3^{k+3}\delta$. Notice that we cannot achieve the statement of the lemma by simply declaring each component to be a separate collection; this separates the components, but there is no bound on the final size since there is no bound on how many components can occur.

Cover X with open sets of diameter less than ϵ, and suppose there are k of them. Choose compact subsets which also cover, and then choose δ so small that the subsets still have diameter less than ϵ when enlarged by $3^{k+3}\delta$. Here "enlarging" a set K by γ means adding all points within distance γ of a point in K.

Now suppose we are given the data of the lemma, with δ the number obtained above. Divide the grope components into k collections, one for each element of the compact cover: the first collection consists of all components whose image intersects the first subset in the cover. The second collection is all components not in the first which intersect the second subset, and so on. If we separate these collections as above, then the choice of δ ensures that the diameter of each entire collection remains less than ϵ.

Finally we separate within each collection. Choose a regular neighborhood for the image of each collection so that the image in X has diameter less than ϵ, and so that distinct neighborhoods are disjoint. Now separate the members of each collection inside the neighborhood (by declaring each member to be a separate collection and using the same procedure). The result satisfies the statement of the lemma.

Noncompact case: We may assume that balls of radius less than 1 in X are compact, and that $3\epsilon < 1$. Cover X locally finitely with compact sets K_i of diameter less than ϵ. For each x in X let $k(x)$ be the number of these sets intersecting the ball of radius 2ϵ about x. Then construct $\delta(x)$ so that for each of the compact sets K_i, the enlargement by $3^{k_i+4}\delta$ has diameter less than ϵ. Here k_i, δ_i denote the maximums of the functions $k(x), d(x)$ over the enlargement of K_i by ϵ.

Now divide grope components into collections and separate as above. There may be infinitely many collections, but the choice of δ prevents any more than k_i from intersecting the ones in collection i during the construction, and ensures the final diameter will be less than ϵ. ∎

Exercise Construct the data required for the lemma from the data of the controlled embedding theorem, 5.4. (ie. construct a properly immersed 3-stage grope of size δ from small disks with transverse spheres, etc. For this, estimate the size growth in the exercise in 2.5). □

2.11 References

The first explicit use of capped surfaces occurs in Freedman and Quinn [1]. There are many implicit occurrences, where they were constructed and then immediately contracted to give spheres, for example in Casson [1] and Freedman [1, lemma 4.2]. The first systematic uses of the Whitney disks arising in a contraction were in Quinn [3] and [5]. However the there presentation was not explicitly in terms of gropes; this formulation was pointed out by R. D. Edwards.

Gropes were introduced by Stan'ko [1], to use in taming wild embeddings in higher dimensions. They were further developed by R. D. Edwards, and J. Cannon, among others (see Cannon [1]). The term "grope" was introduced by Cannon.

The speed and complexity estimates for grope height raising, and the application to killing fundmental group images in 2.9 come from Freedman [4]. The size estimate and the controlled separation lemma comes from Quinn [3].

CHAPTER 3

Capped towers

A capped grope is built of layers of surfaces, and a layer of disks (the caps) at the top. We will need more complicated objects, "towers", whose building blocks are themselves capped gropes. These are defined in the first two sections, and constructed from more primitive data in 3.3. The height of a tower is raised, finitely many times in 3.5 and infinitely in 3.8. The convergent infinite towers of 3.8 provide the raw material for the decomposition theory constructions of the next chapter.

The original proof used somewhat simpler towers, using only accessory disks. Towers of gropes are used here because they allow substantial simplifications in the decomposition arguments.

3.1 Accessory disks

Let $D \subset D^4$ be the standard 2-disk. Define D' by doing a finger move of D through itself. D' then has a pair of intersection points with a canonical embedded Whitney disk W which undoes the finger move (1.7). There is another embedded disk A with boundary on D', passing through exactly one intersection point. This is the model *accessory disk*.

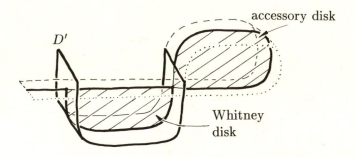

Note that a regular neighborhood of the union $D' \cup A \cup W$ is isotopic rel ∂D to a regular neighborhood of the original embedded disk. Therefore we can think of intersections of things with A and W as representing another sort of "more subtle" intersections with D, (different from those in section 2.1).

Exercise   (Harder than usual) (1) Suppose a surface B intersects W and A. Draw a picture of the corresponding "more subtle" intersections with D, analogous to the picture in 1.2. (2) Find transverse spheres A^t disjoint from D' and W, and W^t disjoint from D' and A. (Hints: See Chapter 12 for pictures. A^t is constructed from 4 copies of W, and W^t from 4 copies of both W and A. Begin with the following picture. Note there are 4 points of intersection $W^t \cup A^t$.) □

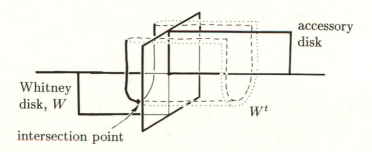

There are also useful transverse capped surfaces, although they require a transverse sphere for D. Example 2 in section 2.1 explains how the linking torus of an intersection point gives a transverse capped surface for a Whitney disk passing through the point. The linking torus of the intersection point not shared with A gives a transverse capped surface disjoint from A. The linking torus of the shared point intersects both. Obtain a transverse capped surface for A by adding a copy of the un-shared torus to the shared one, to remove the intersection with W. Note both of these have the defect that their caps intersect D. To be useful these intersections must be removed, with a transverse sphere.

3.2 Capped towers of capped gropes

A tower has *stories* which are capped gropes. The word "story" is used to refer to layers in a tower to avoid confusion with "stages" in the component gropes. On top of these gropes there is a layer of caps. This means there are two layers of disks at the top: first the caps of the top story capped grope, and then the tower caps.

Formally we define a model for a *capped tower with one story* to be obtained from a model capped grope by introducing standard finger moves in each cap, with disjoint Whitney and accessory disks. The *tower caps* in this case are the Whitney and accessory disks. The *body* is the immersed capped grope (including intersections from the finger moves). If S is a surface we get an S-like tower, by modifying an S-like grope.

Capped towers of more than one story are obtained by replacing the tower caps by capped disk-like towers. As with gropes, a tower is said to have at least k stories if such replacement has been done at least $k - 1$ times. (So there are at least k layers of gropes before the tower caps are reached.) The *body* is then defined to be everything except the tower caps on top of the top story.

As with gropes, immersions of towers are defined by introducing plumbings in the model, and embedding the result. An immersion is *proper* if the plumbings take place only among the tower caps, and the boundary of the lowest surface is mapped to the boundary of the manifold. Notice that the grope caps in a tower do have intersections in a proper immersion, but only the ones already introduced in the model.

Exercise Suppose $A \to M$ is a properly immersed disk-like capped grope, which extends to an immersed one-story tower. Show that every loop in the image of A is contractible in M. □

3.3 Finding towers

An immersion $A \to M$ is said to be π_1-*null* if every loop in the image is contractible in M (ie. $\pi_1(\mathrm{im}A) \to \pi_1 M$ is trivial). Towers will be constructed from gropes with certain π_1-nullity hypotheses. These results are essentially converses to the previous exercise.

Lemma. *Suppose $A \to M$ is a proper immersion of a (union of disk)-like capped grope with height at least three, which is π_1-null and has transverse spheres. Then the embedding of the body of A extends to a proper immersion of a capped tower with one story and arbitrarily many grope stages.*

Recall that A "has transverse spheres" if there is a transverse sphere for each bottom stage component, disjoint from A except for the transverse point.

Proof: The first step is to use the grope height raising proposition 2.7 to extend the body to "arbitrarily many" grope stages, plus one. This modification takes place in a neighborhood of the upper stages, so the new grope is also π_1-null and still has transverse spheres.

Next we produce nice grope caps. Enumerate the components of the top stage, and contract the first one. Push the caps of all the others off the contraction. Then contract the second, and push caps of later ones off. Continue until they are all contracted, to give new disjoint caps for the grope body. Intersections in these new caps automatically come with Whitney disks, since they come from contracting and pushing off.

Choose accessory circles for these Whitney disks. Since these circles are nullhomotopic there are framed immersed disks with them as boundaries. Twist if necessary (as in 1.5) to get the framings to agree with those in the model for accessory disks. The grope together with these Whitney and accessory disks is an immersed one-story tower. The immersion fails to be proper only in that the top caps may intersect lower things.

At each intersection point of a tower caps with either a grope cap or the grope body, push down to the first stage. Add parallel copies of the transverse spheres. Since the transverse spheres are disjoint from the capped grope, except for the standard points, the resulting caps are also disjoint. The immersion is now proper, and satisfies the conclusion of the lemma. ∎

Exercise   Weaken the π_1-null hypothesis in the lemma to: the image of $\pi_1(\ \text{image}A) \to \pi_1 M$ is poly-(finite or cyclic). Weaken the height hypothesis to: height at least 2 on one side, 1 on the other. (See exercise 2.7.) □

The next proposition is the basis for our sharpest conclusion with arbitrary fundamental group. It is also used in the tower height raising procedure in 3.5, although a simpler version would suffice for that. (See the exercise following the proof.)

Proposition. *Suppose A is an immersion of a surface S in M^4, and A^t is a properly immersed π_1-null transverse capped surface for A. Then there is a properly immersed S-like one-story capped tower with arbitrary grope height, transverse spheres, and the same framed boundary as A.*

Here "transverse capped surface" means "transverse 1-stage capped grope" in the sense of 2.6; the caps of A^t may intersect A. Note the π_1-nullity hypothesis applies only to A^t, not to any intersections between A and the A^t caps.

Proof: The plan is to convert A into a capped grope, raising height very carefully. As an induction hypothesis suppose A is a properly immersed capped grope of height at least k with π_1-null transverse capped surface A^t. (Recall from the definition in 2.6 that this means the caps of A^t intersect only the caps of A.) The data given in the proposition is the case $k = 0$. We show how to raise height from k to $k+1$, replicating this data. Then when k reaches 3 we make the upper stages of A π_1-null. The lemma applied to these upper stages then gives the second layer of caps required for a tower.

We begin with an operation which generates transverse spheres. Let B denote a parallel copy of A^t, and contract to get a union of spheres

B'. Define A' and A'^t by pushing intersections with the caps of B off the contraction. B' and A'^t are π_1-null since they are contained in a regular neighborhood of A^t, they are disjoint, and are both transverse objects for A'. The body of A' is the same as the body of A since the caps of B intersect only the caps of A.

The first step is to get the caps of A^t disjoint from A. Some notation is required to see that this can be done preserving the π_1-nullity.

Assume the image of A^t is connected, since components can be joined by a finger move of one cap through another to arrange this. Choose a basepoint in A^t. For each intersection $A \cap (\text{caps of } A^t)$ define an element of $\pi_1 M$ by: go from the basepoint through the image of A^t to the transverse point of intersection with the appropriate component of A. Go through A to the cap intersection with A^t, and return to the basepoint through A^t. This gives a well-defined element of $\pi_1 M$ because A^t is π_1-null. We are concerned with the finite subset of $\pi_1 M$ obtained this way.

Observe that the construction of A', B', A'^t as described at the beginning of the proof increases the number of $A' \cap (\text{caps of } A'^t)$ intersection points, but does not change the subset of $\pi_1 M$. The new A', A'^t intersections arise where they are pushed through each other while pushing off the contraction of B. The group element associated with such an intersection is the product of the one for the AB intersection, and the inverse of the one for the BA^t intersection (see 2.9). Since B is parallel to A^t the first element is already in the subset, and the second is trivial by π_1-nullity.

Now we reduce the π_1-subset by one element. Modify A, A^t as above to obtain transverse spheres B'. Choose an element of the π_1-subset and construct new caps for A^t by: at every point of $A \cap (\text{caps of } A^t)$ with this group element, push the A^t cap down to the bottom stage of A and add copies of B'. This eliminates these intersections, so removes the chosen element from the subset.

To complete the step, observe that the new A^t is still π_1-null. Intersections between the new caps come from intersections in the old caps, selfintersections of one copy of B', and intersections between copies of B'. The π_1-nullity of the original A^t and B' implies that group elements associated to the first two types are trivial. The fact that the curves along which the sums are formed are homotopic (same element of the $\pi_1 M$ subset) implies that the third type are trivial also.

Removing all elements of the π_1 subset this way yields A, A^t with disjoint caps.

The next step is to construct A' by: at each intersection point among caps of A, push one down the other to the bottom stage, and add copies of A^t. This raises height in A by 1, and gives a proper immersion of the result since the caps of A^t are disjoint from A. A^t is a transverse capped surface for A', and is still π_1-null since it is unchanged. Note however that the caps of A^t intersect the caps of A'.

Repeat these steps until the height of A reaches 3, and the caps of A^t are disjoint from those of A. Then, to reduce to a union-of-disks situation, delete a regular neighborhood of a 1-skeleton of the bottom stage. This gives a new manifold with a properly immersed (union of disk)-like capped grope A', still with a π_1-null transverse capped surface. The next step is to modify A' to be π_1-null.

Enumerate the intersection points between A' caps, and suppose there are n of them. Repeat the operation at the beginning of the proof n times to obtain disjoint π_1-null collections of transverse spheres $B_1, \dots, B_n$. Now at the kth intersection point push one sheet down to the bottom stage and add copies of the appropriate sphere in collection B_k. Note that when a sheet is pushed down 3 stages, 8 fingers arrive at the bottom, so 8 copies of the sphere are used. Intersections among these new caps occur only among copies of B_k used in removing a single intersection in the original caps. Since each B_k is π_1-null, and the 8 fingers generated on the way down are homotopic, the intersections in the new caps have trivial associated elements of $\pi_1 M$. Consequently A' with these new caps is π_1-null.

Finally apply the lemma to extend the body of A' to a capped tower. Replacing the regular neighborhood of the 1-skeleton of the bottom stage gives the tower required for the proposition. ∎

Exercise Simplify this proof if the hypothesis is strengthened to: A^t is a π_1-null transverse grope of height at least 4. (Contract one stage of A^t and push off A. Note height can be raised in A^t, and recall the "disjoint near parallels" of 2.4.) ☐

3.4 Transverse gropes

In 2.6 transverse sphere-like capped gropes were constructed for the second stages of a capped grope. The caps of these transverse gropes intersect the caps of the original grope. If the original grope is a stage in a tower this is bad, since the grope caps are part of the body of the tower. The next construction gives transverse gropes which intersect only the tower caps.

Suppose $A \to M$ is an n-story capped tower with grope height at least one in the first story. As in 2.6 divide the second stage surfaces in the first story into $+$ and $-$ sides, and consider the transverse grope E_+^t for the $+$ side. This is constructed from parallel copies of the upper stages of the first story grope, so the caps of E_+^t are parallel to the first story caps. The first story cap intersections have Whitney disk-like gropes; parts of the second story. Each pair of these intersections gives rise to two pairs of intersections between the E_+^t caps and the first story caps. These pairs have Whitney disk-like gropes, parallel to the ones for the original.

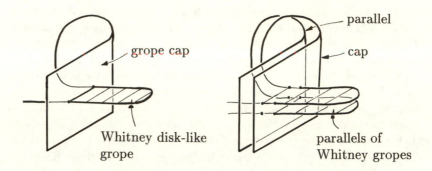

Use these Whitney gropes to do a grope version of the Whitney move: cut out a neighborhood of the Whitney arc on E_+^t, and glue in two copies of the Whitney grope and a parallel of a neighborhood of the Whitney arc on the first story cap. (See the description of the Whitney move in 1.4.)

This makes E_+^t disjoint from the first story caps, but its caps are now parallel copies of the second story caps, and so intersect them. Repeat the process as many times as there are stories.

We draw two useful conclusions. By repeating until the transverse gropes have caps parallel to the top story caps, we get transverse gropes intersecting only the top story caps, and lying in a neighborhood of the body of the tower. These transverse gropes are π_1-null, since components of the body above the first grope stage are π_1-null.

A final application of the procedure—using parallels of the Whitney tower caps—yields transverse gropes disjoint from the body of the tower (except for the standard transverse points). These are not π_1-null.

3.5 Tower height raising

The next result is a "replication" theorem, following the pattern of 2.7. It is stated in general since the proof works in general, but we will only use the simplest case (going from height 1 to height 2).

Proposition. *Suppose $A \to M$ is a proper immersion of an n-story capped tower with grope height at least 4 in the first story. Then the* *top story grope caps can be changed so the body extends to a properly immersed $(n + 1)$-story capped tower with arbitrarily grope height in the top story.*

Proof: The construction in broad outline is like that of 2.7: we push intersections among the tower caps down and add copies of transverse sphere gropes.

Begin by constructing transverse gropes E_+^t and E_+^t as in 3.4 whose caps intersect only the tower caps. Contract these and push off the tower caps, to get transverse spheres disjoint from the tower except for the standard points. Delete a neighborhood of the first stage, and denote the resulting 4-manifold by M'. The higher stages now constitute a (union of disk)-like capped tower A' in M', which has transverse spheres. These transverse spheres will be used as in 3.3 to convert nullhomotopies into tower caps.

Next construct transverse gropes E_+^t for surfaces in the second stage of A', which intersect only the top story grope caps on the $-$ side, and lie in a regular neighborhood of the body. Contract the top stage of E_+^t, and push the $-$ side caps off. Both remain π_1-null since this takes place in the neighborhood of the body. Note the $+$ side is unchanged.

The next step is to raise height in E_+^t to get "arbitrary" height, plus 1. We can construct many disjoint transverse gropes by repeatedly contracting a parallel of E_+^t, contracting the top stage, and pushing the caps of E_+^t off. (These are "disjoint near parallels" as in 2.4; see also the beginning of the proof of proposition 3.3.) These gropes have "arbitrary" height, and are all π_1-null.

Now push all intersections among the tower caps on the $+$ side down to the second stage, getting—as in 2.5—a cascade of fingers coming down. Add a separate disjoint near parallel of E_+^t to each one of these fingers, to make it disjoint from the grope. This converts the $+$ tower caps into properly immersed capped gropes, which will be the body of the new story.

We now have an uncapped tower, which we denote by T: the original $-$ side with slightly modified top story caps, the $+$ side, and the capped gropes in the new story on the $+$ side. T is π_1-null, on the $-$ side because that side is in a neighborhood of the original side, and on the $+$ side because top cap intersections come from single disjoint near parallels of E_+^t, and these are all π_1-null.

Different top story caps in T may intersect. To fix this raise grope height in the top story by 1, contract the top surface components one at a time and push the others off. The result is still π_1-null because this operation takes place in a regular neighborhood.

Next construct tower caps for T, as in lemma 3.3. Since the top story caps were obtained by contraction, there are Whitney disks for all the intersections. Use the π_1-nullity to find accessory disks. Then push intersections of these disks with the body down to the bottom stage and add copies of the transverse spheres constructed at the beginning of the proof. This yields tower caps disjoint from the body, so a properly immersed capped tower.

This construction raises tower height on the $+$ side. To complete the proof raise height on the $-$ side by switching $+$ and $-$ in the argument above. ∎

3.6 Controlled towers

The next proposition continues the controlled material of 2.10, and is required in the proof of the controlled embedding theorem. The objective is to manipulate towers with control on their diameters. The main step is to obtain 1-story towers with small disjoint components, since further constructions take place inside regular neighborhoods of this data.

Proposition. *Suppose X is a locally compact metric space, and $\epsilon :$ $X \to (0, \infty)$ is continuous. Then there is $\delta : X \to (0, \infty)$ such that if $M^4 \to X$ is given, $A \to M$ is a (possibly infinite) δ-π_1-null proper (union of disk)-like grope with height at least 3 and transverse spheres of diameter $< \delta$, and the images in M are locally finite, then the body extends to a 1-story capped tower with disjoint components of diameter $< \epsilon$.*

Here δ-π_1-null means that loops in the image of the grope extend to maps of 2-disks of diameter less than δ. Note that it follows that components of the image of A have diameter less than δ.

Proof: The first step is to get nice grope caps, as in the proof of 3.3. Enumerate the caps, raise height to convert them into properly immersed gropes, and contract them one at a time in order, pushing all others off. This process takes place in (arbitrarily small) regular neighborhoods of the image of A. Since the δ-π_1-null condition implies that components of the image of A have diameters less than δ, this can be done without increasing diameter.

Intersections in these caps have Whitney disks, inside the regular neighborhoods so of diameter $< \delta$. Choose accessory circles, and extend to immersed accessory disks of diameter $< \delta$ using the δ-π_1-null condition. Push intersections of these disks with the grope down to the first stage, and add copies of the transverse spheres. This yields a properly immersed capped tower, and if $\gamma_0 : X \to (0, \infty)$ is given then for small enough δ each component of the tower will have image in X

of diameter $< \gamma_0$. (If δ is constant then the accessory disks may have diameter 3δ, so component images will have diameter $< 7\delta$. Therefore $\delta = (1/7)\gamma_0$ will work. Existence in the function case follows from the constant case, but the relationship is not so tidy.)

The problem with these towers is that tower caps of distinct components may intersect. To separate them we convert them into gropes and use the controlled grope separation lemma 2.10.

Collect the second stage surfaces in the grope into $+$ and $-$ sides, and construct transverse gropes E_+^t as in 3.4 which are disjoint from the grope but which may intersect the tower caps. Push tower cap intersections with the $+$ side caps down to the second stage and add copies of the E_+^t. This converts the caps into (second story) immersed gropes; the $+$ caps are proper but the $-$ caps still have the original intersections in the first stage. Contract the $-$ side gropes completely, and push off the $+$ side. This increases diameter by a factor of 7^2 in the constant case, and in general if γ_1 is given then the resulting diameters are $< \gamma_1$ if γ_0 is small enough.

Now apply the separation lemma to see that if γ_2 is given then if γ_1 is small enough the caps of the $+$ side second story gropes can be changed to be mutually disjoint, and with diameter $< \gamma_2$.

Next reverse $+$ and $-$. Construct transverse gropes E_-^t from parallels of the $+$ side, intersecting only the caps of the $+$ side second story gropes. Completely contract the $+$ side gropes and push off the E_-^t. We are back to tower caps on the $+$ side, but they are mutually disjoint and disjoint from the E_-^t. Push down intersections among the $-$ side tower caps and add copies of the E_-^t. This gives properly immersed second story gropes on the $-$ side. Apply the separation lemma in a neighborhood of the image of these gropes to find new, disjoint, caps. Contract to get disjoint tower caps on the $-$ side. If γ_2 is small enough then the resulting towers will have diameters $< \epsilon$, so satisfy the conclusions of the proposition. ∎

3.7 Squeezed towers

The tower height raising argument of 3.5 uses a great many parallel copies of the first story, so produces a second story which winds tightly around the first. The objective here, in preparation for the convergence arguments of the next section, is to pull the second story away from the first and reduce its size. Size conditions are used with rather different intent and effect than the controlled material of 3.6; sizes are measured in M itself.

Lemma. *Suppose $A \to M$ is a proper immersion of a one-story capped tower with grope height at least 4, and $\epsilon > 0$. Then the embedding of*

*the grope body extends to a properly immersed two-story capped tower
with arbitrarily many stages in the second story, and such that distinct
components of the upper story tower lie in disjoint balls of radius less
than ϵ.*

Here "upper story tower" means the one-stage tower formed from the
second-story grope and the tower caps.

Proof: The first step is to extend the grope body to a 2-story tower
satisfying: there are "arbitrarily many" stages in the second story, distinct second story components have disjoint images, and the upper story
tower is π_1-null in the complement of the first story.

To achieve this, raise the tower height to 3 with two applications of 3.5.
Then discard the tower caps and totally contract the third story gropes.
The resulting disks are tower caps for the first two stories. Distinct
caps are disjoint because the third story grope components are. Finally
the original tower caps (above the third story) give nullhomotopies for
generating loops for the fundamental group of the image (see exercise
3.2). The nullhomotopies are disjoint from the first story grope because
the tower caps are.

Next observe that a regular neighborhood of a tower is also a regular
neighborhood of a 1-complex. The lowest stage collapses to the attaching
curves of the next stage up, union with embedded arcs joining attaching
curves lying in the same component. Iterating this observation collapses
the tower to the top caps union trees. Then the top caps collapse to loops
passing through intersection points, and more arcs. The analogous fact
for capped surfaces is mentioned in 2.2.

The boundary of the regular neighborhood intersects the first story
tower caps in circles. As a pair the neighborhood with these circles is a
relative regular neighborhood of the mapping cylinder of the circles to
the 1-complex.

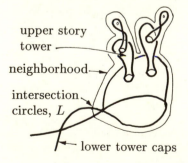

upper story
tower

neighborhood

intersection
circles, L

lower tower caps

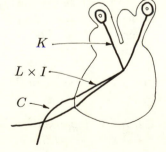

K

$L \times I$

C

We fix some notation: let K denote the 1-complex, L the intersection of the boundary of the neighborhood with the grope caps, and $L \times I$ the mapping cylinder. C denotes the part of the grope caps outside the neighborhood. So for example $C \cup (L \times I) \cup K$ has regular neighborhood isotopic to a neighborhood of the grope caps union the upper story towers. The end $L \times \{1\}$ of the mapping cylinder is identified with its image in K, but the notation does not reflect this explicitly.

The next step is to move the upper story towers into small balls by isotopy. Since these towers lie in a regular neighborhood of K we can do this by moving K. This can be done because K is a nullhomotopic 1-complex, and homotopy implies isotopy for 1-complexes.

In more detail, choose balls disjoint from the tower and each other, of radius less than ϵ. Choose one for each component of the upper story tower. The π_1-nullity of the tower implies that K is nullhomotopic in the complement of the first story. Choose a homotopy of each component of K to the center of the associated ball.

This defines a map $K \times I \to M$. Approximate it, rel $K \times \{0\}$ and keeping $K \times \{1\}$ in the balls, to be an immersion in general position. It intersects itself, the grope caps, and the mapping cylinder $L \times I$ in points. Push all these surfaces off the $K \times \{1\}$ end of $K \times I$, by disjoint finger moves. This changes the tower by isotopy, and makes the map of $K \times I$ an embedding disjoint from the grope body, the interior of $L \times I$, C, and with $K \times \{1\}$ still inside the small balls.

Use the embedding of $K \times \{1\}$ to define an ambient isotopy of a neighborhood of K into the balls, by pushing in the I coordinate. We may assume the neighborhood contains the upper story towers, and that the isotopy is fixed on C and the bottom story grope body. The image of the tower under this isotopy therefore satisfies the conclusions of the lemma. ∎

3.8 Convergent infinite towers

Proper immersions of towers are embeddings except for intersections among the tower caps. The idea here is that if we raise tower height infinitely then there will be no tower caps, so no extra intersections. For this to be useful the infinite construction must converge in an appropriate sense, which is arranged with the tower squeezing lemma 3.7.

Model capped towers are described by an iterative construction. To be precise, suppose we have a model capped tower with n stories. Choose disjoint 4-balls in the model which intersect only the top caps, and these in standard $D^2 \subset D^4$. Replace these D^2, D^4 pairs by model capped towers. This gives a model capped tower with $n + 1$ stories. Finally we require that further modifications—introducing more stories—take place inside the balls used for this step.

We say that an infinite sequence of such modifications *converges* provided for each $\epsilon > 0$ there is n such that all the balls used to modify story n have diameter less than ϵ. The nesting condition ensures that there is a well-defined image for a convergent sequence of modifications. We refer to this image as a *convergent infinite tower*.

A convergent tower has a limit set; limit of the nth story, as $n \to \infty$. This is almost always a Cantor set.

Exercise Suppose at each step in the construction of a model infinite tower each cap is replaced by a nontrivial (ie. $\neq D^2$) capped tower. Show that the limit set is a Cantor set, by showing that it is compact, completely disconnected, and has no isolated points. □

The next objective is to define *pinched regular neighborhoods* of the convergent tower, pinched at the limit set. The point here is that embeddings of towers, etc. have been defined as embeddings of neighborhoods in the model rather than just the 2-complex spines. However we will not be able to embed a full neighborhood of an infinite tower, so "embedded convergent tower" will mean an embedding of a pinched neighborhood.

Suppose we have a regular neighborhood of a model n story tower. Modify it so that it intersects the balls used to introduce the $n+1$ story in a standard regular neighborhood of $D^2 \subset D^4$. Then replace this by a regular neighborhood of the new story inside the ball. We require future modifications to the neighborhood to take place inside these balls. This process also converges, and the result is a pinched neighborhood.

Abstractly we can recognize a convergent tower together with its limit points as the endpoint, or Freudenthal, compactification of the tower. The pinched regular neighborhood is similarly the endpoint compactification of a regular neighborhood of the tower.

If X is a locally compact space the *endpoint compactification* $E(X)$ is defined as follows: $E(X) - X$ consists of equivalence classes of decreasing sequences of open sets U_i. Each U_i is a nonempty component of the complement of some compact set in X. We require also that if K is compact in X then for some i, $X - K \subset U_i$. Two sequences U, U' are equivalent if for each i there is j such that $U_i \supset U'_j$ and $U'_i \supset U_j$. Finally the topology of $E(X)$ is generated by open sets in X and sets $V \subset E(X)$ so that $V \cap X$ is a component of the complement of a compact set, and $V \cap (E(X) - X)$ consists of all sequences which are eventually inside $V \cap X$.

If X is hausdorff, connected, locally connected, and locally compact, then $E(X)$ is compact hausdorff. In particular this is the case with our infinite towers.

Proposition. *Suppose $A \to M$ is a properly immersed 1-story capped tower with grope height at least 4, and $\{n_i\}$ is a sequence of positive integers. Then the embedding of the grope body extends to an embedding of a pinched regular neighborhood of a convergent tower, with grope height n_i in story i.*

Proof: This follows easily from the squeezing lemma, 3.7. Apply the lemma to get a two-story tower with second story in balls of radius $< 1/2$. Delete an open regular neighborhood of the first story, and take regular neighborhoods of components of the upper story tower contained in the given balls. These components inside their neighborhoods satisfy all the hypotheses of the squeezing lemma, so this construction can be repeated inside the neighborhood. Repeat infinitely many times, requiring that in the ith repetition the balls all have radius less than $1/i$. Then regular neighborhoods chosen as in the construction of the model give the required pinched regular neighborhood. ∎

Exercise Define a convergent infinite *grope* by iterating the operation of 2.1. Show that in the situation of the proposition there is a convergent infinite grope in M. □

3.9 References

Accessory disks, as companions to Whitney disks, were introduced in Freedman and Quinn [**2**], and further developed in Quinn [**4**].

The basic idea of constructing towers, particularly infinite ones, is due to A. Casson [**1**]. His towers were composed entirely of disks, all "accessory" in the sense that the attaching circle goes through exactly one intersection point in the previous stage. Framing conditions were given in terms of link diagram descriptions of the boundary of a regular neighborhood, rather than directly by a model as in 3.1. Finally there was no convergence condition on his infinite towers. Other expositions of height raising for Casson-type towers are given in Gompf and Singh [**1**], and Quinn [**5**].

Convergence conditions were obtained by Freedman [**1**], [**2**] for Casson's towers, using an analog of 3.5. The derivation of the analog of 3.7 from 3.5 follows a suggestion of R. Edwards.

Towers of Whitney and accessory disks (towers of capped gropes of height 0 in the terminology here) were developed by Quinn [**5**]. This treatment also used models rather than link diagrams. Towers with one story ("twice capped gropes") were used in Freedman [**4**]. General towers of capped gropes appear here for the first time.

The endpoint compactification was introduced by Freudenthal [**1**].

CHAPTER 4

Parameterization of convergent towers

It follows from section 3.8 that when the hypotheses of the main theorem are satisfied then the manifold has a subset homeomorphic to a certain subset of D^4, namely a pinched regular neighborhood of a convergent infinite tower. The objective of this chapter is to show that this pinched neighborhood contains a correctly framed embedded disk.

4.1. The parameterization theorem

We establish some notation to be used throughout this chapter. $C \subset D^2 \times D^2$ is a pinched regular neighborhood of a disk-like convergent tower, as described in 3.8. $C \cap \partial(D^2 \times D^2)$ is arranged to be $S^1 \times D^2$, and is denoted $\partial_0 C$. $\partial_1 C$ is the frontier of C; $C \cap$ (closure of $(D^2 \times D^2 - C)$). In these terms we can be more precise about the objective.

Theorem. *Suppose C is a pinched regular neighborhood of a connected convergent infinite tower with grope height at least 3 in the first story. Then there is another such neighborhood of a tower $C' \subset C$ and a homeomorphism $D^2 \times D^2 \to C'$ which is the identity on $S^1 \times D^2$.*

The construction of such a homeomorphism is rather indirect. In 4.2 we construct a map $D^2 \times D^2 \to C'$ which is not a homeomorphism. This is used in 4.3 to construct a homeomorphism from a subset of $D^2 \times D^2$ into C'. In 4.4 maps to a common quotient $\alpha : D^2 \times D^2 \to Q$ and $\beta : C' \to Q$ are defined, roughly by dividing out the complement of the partial parameterizations of 4.3. These maps α, β are shown to be approximated by homeomorphisms in 4.5 and 4.6 respectively. The composition of these homeomorphisms gives the homeomorphism required for the theorem.

4.2. Singular parameterization

Proposition. *Suppose C is a pinched regular neighborhood of a convergent infinite tower with grope height at least 3 in the first story. Then there is another such neighborhood of a tower C' and a map $D^2 \times D^2 \to C'$ which is onto, is the identity on $S^1 \times D^2$, and whose* *point inverses are points, and arcs which are fibers in a ("wild") collar of $D^2 \times S^1$.*

A collar here means a map $D^2 \times S^1 \times I \rightarrow D^2 \times D^2$ which is the inclusion on $D^2 \times S^1 \times \{1\}$ and is a homeomorphism on its image. Such a collar is *wild* if it cannot be extended to a collar which contains it in the interior. Alternatively it is wild if the image of $D^2 \times S^1 \times \{0\}$ is not collared on the inside. This parameterization does not imply that C' itself is a disk. For example the Alexander horned sphere divides the 3-sphere into a 3-disk and a nonsimply-connected object with boundary S^2. The nonsimply-connected object has a 3-dimensional analog of the parameterization of 4.2.

Note that the map of 4.2 restricts to a homeomorphism $D^2 \times S^1 \rightarrow \partial_1 C'$.

Proof of the Proposition: We need a dual view of the model for a (disk-like) capped tower. The spine of the model is a 2-complex in $D^2 \times D^2$ which has all of $D^2 \times D^2$ as a regular neighborhood. A regular neighborhood of the body is therefore obtained by deleting certain 2-handles, the duals of the caps, from $D^2 \times D^2$. By a 2-handle we mean here an embedding $(D^2 \times D^2, D^2 \times S^1) \rightarrow (D^2 \times D^2, D^2 \times S^1)$. If we raise height in the model by replacing the caps by towers inside the dual handles, then we get new 2-handles dual to the new caps, inside the previous 2-handles.

Now consider regular neighborhoods of a model infinite tower. The dual description of the finite stories gives a nested infinite sequence of families of 2-handles in $D^2 \times D^2$, say $(H_{i,*})$. A regular neighborhood of the body of stage i is the complement of the 2-handles in collection i; $D^2 \times D^2 - \bigcup_* H_{i,*}$. Thus a regular neighborhood of the infinite tower (not including limit points) is the complement of the intersection of the sequence; $D^2 \times D^2 - \bigcap_i (\bigcup_* H_{i,*})$.

Next recall that a pinched regular neighborhood of a convergent tower is homeomorphic to the endpoint compactification of a regular neighborhood of the body (3.8). This compactification is just the quotient of $D^2 \times D^2$ obtained by identifying components of the intersection to points. The quotient map therefore gives a map $D^2 \times D^2 \rightarrow C$ which is onto, the identity on $S^1 \times D^2$, and has point inverses points and components of the intersection of the families of 2-handles. This is valid for any convergent tower C. What we have to show is that a new tower C' can be constructed and dual 2-handles in the model chosen so that components of the intersection are arcs as described in 4.2.

Let $d: D^2 \times S^1 \times I \rightarrow D^2 \times D^2$ be the collar of $D^2 \times S^1$ which is radial in the second coordinate and has radius 1/2 (so the image is the complement of $D^2 \times (\text{int} \frac{1}{2} D^2)$. We construct a convergent tower $C' \subset C$ with sequence of dual 2-handles $H_{i,*}$ in the model which satisfy:

(1) there are collars $c_i : D^2 \times S^1 \times I \to D^2 \times D^2$ taking $D^2 \times S^1 \times \{1\}$ to $D^2 \times S^1$, such that $H_{i,m}$ intersected with c_i is the standard collar d on $\partial_0 H_{i,m}$,

(2) $c_{i+1} = c_i$ except on $\bigcup_* (\partial H_{i,*}) \times [0, (\frac{1}{2})^i]$, and $c_{i+1}(\partial_0 H_{i,m} \times I) \subset H_{i,m}$, and

(3) the diameters of $\partial_0 H_{i,m}$ and $H_{i,m}(D^2 \times (\frac{1}{2} + (\frac{1}{2})^i)D^2)$ are less than $(1/2)^i$.

Such a tower has the desired properties. Define K to be $\bigcap_i H_{i,*}$, and $\partial_0 K$ to be the intersection with $D^2 \times S^1$. The size condition (3) implies that components of $\partial_0 K$ are points, each given by the intersection of a nested sequence $\partial_0 H_{i,m(i)}$. Condition (2) defines a limit collar c, except on $\partial_0 K \times \{1\}$. The size and containment conditions in (2) and (3) imply that c is defined at least as a map on these points also. The map is an embedding because the point $c(\bigcap_i \partial_0 H_{i,m(i)} \times \{0\})$ is contained in the set $\bigcap_i H_{i,m(i)}(D^2 \times (\frac{1}{2} + (\frac{1}{2})^i)D^2)$, and for different sequences these sets are disjoint from each other and the rest of the image of c. Finally condition (1) and the size conditions imply that the intersection of a nested sequence $\bigcap_i H_{i,m(i)}$ is the fiber arc $c(\bigcap_i \partial_0 H_{i,m(i)} \times I)$.

Towers satisfying conditions (1)-(3) are constructed by induction, using the following lemma.

Lemma. *If $\epsilon > 0$ then there is k such that the model for any one-story capped tower with grope height at least k is isotopic to one with dual 2-handles whose boundary solid tori lie in ϵ balls about points in $\{0\} \times S^1$.*

Proof of the lemma: If a capped surface is obtained by one 0-surgery on D^2 (see 2.1) then the dual 2-handles consist of (regular neighborhoods of) the following pair of disks.

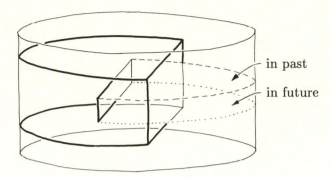

These disks were described in 2.1 as the "more subtle" intersections of surfaces with $D^2 \times D^2$. A capped surface obtained by j 0-surgeries has dual 2-handles obtained by simply stacking up j copies of this picture.

A grope of height 2 is obtained by replacing each cap with a capped surface inside the dual 2-handle. Dually each 2-handle is replaced by a stack of "linking" subdisks, in copies of the same picture. For an k-stage grope make such replacements k times.

The picture in the boundary is to pass from a solid torus $D^2 \times S^1$ to two linked solid tori inside it (or a stack of copies of this). This process is called *Bing doubling*.

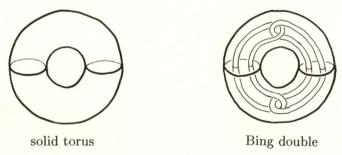

solid torus   Bing double

Since these solid tori can be compressed arbitrarily closely to $\{0\} \times S^1$, to get them within ϵ of points of the form $(0, \theta)$ it is sufficient to arrange the projections to S^1 to have diameter less than ϵ.

Claim: Suppose $i \colon S^1 \to D^2 \times S^1$ is a smooth embedding which on half the circle is a monotone loop of length less than $r \geq \epsilon$, and on the other half has diameter less than ϵ, measured in the S^1 coordinate. Then a Bing doubling in a small neighborhood can be arranged to have components which are monotone loops of length less than $r - \epsilon$ on half the circle and have diameter less than ϵ on the other half.

By a "monotone loop" we mean that the projection to S^1 goes monotonically a distance r, and then goes monotonically back.

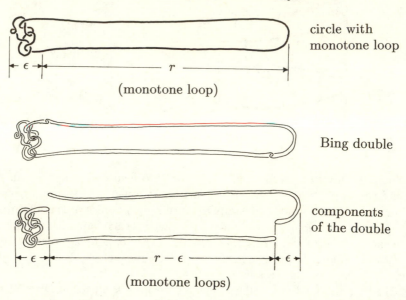

circle with
monotone loop

(monotone loop)

Bing double

components
of the double

(monotone loops)

The pictures indicate the proof of the claim. Now note that the first doubling of the center circle of $D^2 \times S^1$ yields circles which have a piece of diameter ϵ and a monotone loop of length less than π (1/2 of the length of S^1). Therefore Bing doubling and applying the claim k more times, where $k\epsilon > \pi$, gives circles of diameter less than ϵ.

Regular neighborhoods of these circles are the boundary solid tori of the dual 2-handles to the *grope caps* of a capped grope of height k. To complete the lemma we must introduce finger moves in these, and pass to the duals to the tower caps. As with adding a grope stage, these duals are inside the previous ones. But these automatically satisfy the size conditions since the previous ones already do. ∎

Completion of the Proposition: Suppose as induction hypothesis we have a capped tower in C with $n+1$ stories, grope height at least 3 in the top story, and such that components of story i have diameter less than $1/i$. Let $J_i \subset D^2 \times D^2$ be a model for the first i stories ($i \le n$), with 2-handles $H_{i,*}$ dual to the tower caps. J_{i+1} is obtained by replacing the tower caps of J_i by one-story capped towers inside the dual handles. Suppose further that the dual handles satisfy conditions (1)-(3) for $i \le n$.

Consider one of the dual handles to story n; $H_{n,m}: D^2 \times D^2 \to D^2 \times D^2$. By uniform continuity there is $\delta > 0$ so that a set of diameter less than

δ has image of diameter less than $(1/2)^{n+1}$. According to the lemma there is k so that if the tower caps in J_n are replaced by any capped tower of grope height at least k then there are duals to the caps which have boundary solid tori of diameter less than δ. Since there are finitely many of these handles we conclude that there is a k which works for all of them.

In C we have a tower with $n+1$ stories. Apply 3.5 and 3.7 to replace the top story by a 2-story capped tower with at least k stages in the first story, at least four in the second, and so that the image of components are appropriately small. Define J_{n+1} to be the model for the first $n+1$ stories. Then by choice of k we can find dual handles with small boundary tori. Denote these by $H_{n+1,*}$.

We can arrange that the collars on ∂_0 match up in H_n and H_{n+1}, on $[(\frac{1}{2})^n, 1]$. Specifically, $H_{n+1,*}(d(y,t)) = H_{n,*}(d(\partial_0 H_{n+1}(y), t))$ for $y \in D^2 \times S^1$ and $t \in [(\frac{1}{2})^n, 1]$. By compressing toward the centers of $H_{n,*}$ we can arrange that each $H_{n+1,*}(D^2 \times (\frac{1}{2} + (\frac{1}{2})^{n+1})D^2)$ has diameter less than δ. Finally we can modify the standard collar on $(\partial_0 H_{n,*}) \times [0, (\frac{1}{2})^n]$ to agree with the collars in $H_{n+1,*}$.

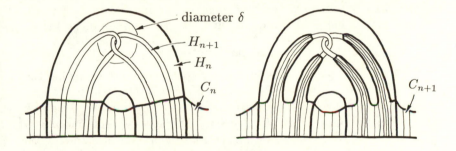

The modified collar is c_{n+1}, and it together with the handles $H_{n+1,*}$ satisfy conditions (1)-(3) for $i \leq n+1$. This completes the induction step. ∎

Exercise Show that the boundary solid tori of the dual handles of tower caps are regular neighborhoods of boundary circles of immersed 2-disks in $D^2 \times S^1$. These disks lie inside the boundary solid tori of the dual handles of the top grope stage, and intersect in disjoint arcs (there are no triple points). This can be done two ways: (1a) Note the duals of caps for D^2 with one finger move are regular neighborhoods of the disks in the following picture:

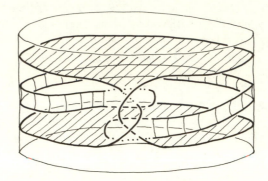

This picture is the solution to exercise (1) in 3.1. To get the immersed 2-disks deform the disks shown rel boundary into $\partial D^2 \times I$.

(1b) Show that the core circles of the solid tori are nullhomotopic, in the tori for the top grope stage. Then show that nullhomotopic embedded circles in a 3-manifold bound immersed disks whose intersections and self-intersections are disjoint arcs. □

Exercise Show that any convergent infinite *grope* has a singular parameterization as in 4.2. (In particular it is unnecessary to pass to another convergent grope inside the first. See the exercise in 3.8.) □

4.3. Partial parameterization

We avoid the singular arcs in 4.2, but at the expense of having the parameterization only partially defined. Subsets of $D^2 \times D^2$ are deleted to cut the arcs down to Cantor sets, and then (roughly speaking) we spread out the points in the Cantor sets. This gives a homeomorphism of a subset of $D^2 \times D^2$ into a pinched neighborhood of a tower. To describe the subset we need some notation.

Let Ψ denote the set of finite sequences with entries "0" or "2". If $s \in \Psi$ then ds is the sequence obtained by deleting the last entry (if there is one; for the empty sequence define $d\phi = \phi$). The length of the sequence is denoted $\ell(s)$. A sequence of length i with all entries "2" is denoted $i : 2$.

We will need a description of the standard Cantor set in terms of Ψ. If $s \in \Psi$ then $.s$ specifies the base 3 expansion of a number in $[0, 1]$; first a decimal point, then the sequence of "0" and "2" specified by s. Define J_s to be the closed interval $[.s1, .s2]$, were $.s1$ means the base 3 expansion specified by s, followed by a "1". The complement $[0, 1] - \bigcup_{s \in \Psi} \text{int} J_s$ is then exactly the usual middle thirds description of the Cantor set; the points in I with a base 3 expansion which does not contain a "1".

In the proposition $D^2 \times S^1 \times [0,1]$ denotes a radial collar of $D^2 \times S^1 \subset D^2 \times D^2$. B denotes an open ball in the interior of $D^2 \times D^2$ obtained as the complement of this collar and a thin collar on $S^1 \times D^2$.

Proposition. *Suppose C' has a singular parameterization as constructed in 4.2. Then there are finite collections $T_{s,*}$ for each $s \in \Psi$ of disjoint solid tori in $D^2 \times S^1$ satisfying:*

(1) *$T_\phi = S^2 \times S^1$, and if $s \neq \phi$, $T_{s,*}$ are regular neighborhoods of boundaries of immersed 2-disks in $\text{int} T_{ds,*}$, which intersect in disjoint arcs (no triple points),*

(2) *each component of $T_{s,*}$ has diameter less than $(1/2)^{\ell(s)}$, and*

(3) *there is a map $D^2 \times D^2 - B - \bigcup \text{int}(T_{s,*} \times J_s) \to C'$ which is the identity on $S^1 \times D^2$, takes $D^2 \times S^1$ to $\partial_1 C'$, is a homeomorphism onto its image, and $T_{ds,*} \times J_s$ has image in C' of diameter less than $(1/2)^{\ell(ds)}$.*

Proof: We denote the set $D^2 \times D^2 - \bigcup \text{int}(T_{s,*} \times J_s)$ by E, and begin by giving a union description of E. The interval J_s is the middle third of a larger interval. Denote the upper and lower thirds of this interval by J_s^+ and J_s^- respectively. Explicitly, $J_s^- = [.s, .s1]$ and $J_s^+ = [.s2, .s(o:2)]$. In constructing E by deleting middle thirds, note that the upper and lower thirds are left in E. Therefore E is the closure of $\bigcup_s((T_{s,*} - T_{s2,*}) \times J_s^+ \cup (T_{s,*} - T_{s0,*}) \times J_s^-)$.

The limit points in the closure are intersections of nested sets $T_{r(j),w(j)} \times J_{r,j}$. Here r is an infinite sequence of "0" and "2", $r(j) \in \Psi$ denotes the first j terms, $T_{r(j),w(j)}$ is a nested sequence of the solid tori, and $J_{r,j}$ denotes $J_{r(j)}^-$ or $J_{r(j)}^+$ if the $j+1$ entry of r is "0" or "2" respectively.

The picture shows a cross section of E from ∂_0 to one of the singular arcs $\{y\} \times [0,1]$, where $y \in \bigcap_s T_{s,*}$.

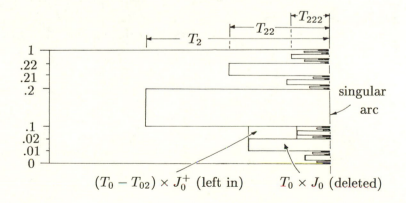

Notice that if tori $T_{s,*}$ are defined for some subset $\Phi \subset \Psi$ which is closed under d, then we can define a subset $E(\Phi)$ by taking the union over Φ and taking the closure.

As a first approximation we see how close we can come to proving the proposition using 4.2 directly. Recall that there are nested families of dual 2-handles $H_{n,*}$ (duals to caps for story n) with the property that they intersect the collar in products, diameters of the boundary solid tori go to zero, and diameters of the images in C' go to zero. Choose $n(i)$ such that each component of $H_{n(i),*}$ has image in C' and boundary solid torus both of diameter less than $(1/2)^i$. Require also that $n(i+1) \geq n(i) + 2$. For $s \in \Psi$ define $T_{s,*}$ to be the boundary tori of the handles $H_{n(\ell(s)),*}$. (So the collections depend only on the length of the sequences).

These tori satisfy (2) by choice of $n(i)$. The first exercise in 4.2 gives the 2-disks needed for (1). The map in 4.2 defines a map as in (3) which fails only to be one-to-one. It is not one-to-one on the intersection of E with fibers of the collar which correspond to intersections of all the families of solid tori. At these fibers we have deleted all the middle thirds from E, so these give point inverses in E which are Cantor sets. The problem is to separate the points in these Cantor sets.

We now start assembling the pieces of the proposition. Let Φ_m denote the subset of Ψ of sequences with exactly m "0" entries. We will define subsets $E(\Phi_m)$ and maps $q_m \colon E(\Phi_m) \to C'$ by induction on m.

Let $p_0 \colon D^2 \times D^2 \to C'$ denote the singular parameterization of 4.2. For $s = i : 2$ in Φ_0 let $T_{s,*}$ be the boundary tori of handles $H_{n(i),*}$ chosen as above. This defines $E(\Phi_0)$ and we define $q_0 \colon E(\Phi_0) \to C'$ to be the restriction of p_0. Since the intervals used are the upper thirds $[.s2, 1]$ this subset contains $D^2 \times S^1 \times \{1\}$ and its closure intersects each singular arc in only one point. Therefore q_0 is one-to-one.

We set up some notation in preparation for the next step in the construction. Certain dual handles $H_{n(i),*}$ were chosen above, and for $s = i : 2$, $T_{s,*}$ was defined to be the intersection of these with the boundary. Define $H'_{s,*}$ to be $H_{n(i),*} - T_{s,*} \times (.s1, 1]$. This is a union of copies of $D^2 \times D^2$ with collars on ∂_1 given by $T_{s,*} \times J_s^-$. The picture shows a cross section of $E(\Phi_0)$ and one of the H', similar to the section of E above.

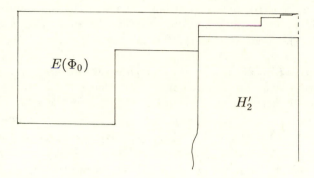

Now (still with $s = i : 2$) note that the difference $H'_{ds,*} - H'_{s,*}$ is a neighborhood of an uncapped, unconnected tower with $n(i) - n(i-1) \geq 2$ stories and ∂_0 equal to $\partial_0 H'_{ds,*}$. The map p_0 is a homeomorphism on this difference, so we get a copy of this tower in C'. By contracting the upper stories we get a capped tower with one story in C'.

Apply 3.8 and 4.2 to see that each component of the one-story tower contains a convergent infinite tower parameterized as in 4.2. More precisely there are maps $p_{1,s}: H'_{s,*} \to C'$ which extend p_0 on $\partial_0 H'_{s,*}$, and are singular parameterizations of their images. Repeat the construction above in the components of this parameterization: choose integers so that boundaries and p_1 images of the dual handles are appropriately small, and define these boundaries to be the $T_{t,*}$ for $t \in \Phi_1$.

We describe the indexing more precisely. Suppose T is a torus boundary of a handle $\hat{H}_{m(j),*}$, in the tower constructed inside $H'_{i:2,*} - H'_{(i+1):2,*}$. Then the index on T is $(i:2)0(j:2)$. Tori are now defined for sequences with a single "0" entry, so the subset $E(\Phi_1)$ is defined. The map q_0 is extended to q_1 on $E(\Phi_1)$ by the singular parameterizations $p_{1,s}$.

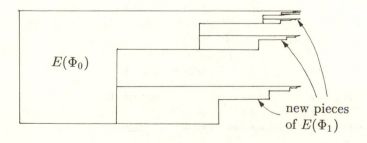

Repeat this construction inside the segments of the inner towers. This gives solid tori with indices which contain two "0" entries, and defines q_2 on $E(\Phi_2)$. Persistent repetition gives tori satisfying (1) and (2), for all indices Φ_m, and a map $q_\infty \colon \bigcup_m E(\Phi_1) \to C'$ satisfying the size restrictions of (3). The size restrictions imply that this extends to a map of the closure $q \colon E \to C'$. It remains to see that q is one-to-one.

First we see that q_∞ is one-to-one. Recall that the s, w component of the image of $E(\Phi_1) - E(\Phi_0)$ (for $s \in \Phi_0$) was constructed inside $p_{1,s}(H'_{s,w})$ which is in $p_0\bigl(H'_{s,w} - \bigcup_{\ell(t) > \ell(s)}(H'_{t,*})\bigr)$. These sets are mutually disjoint, and are disjoint from $q_0\bigl(E(\Phi_0)\bigr)$, so images of components of $E(\Phi_1) - E(\Phi_0)$ are disjoint from each other and the image of $E(\Phi_0)$. Since q_0 was observed to be one-to-one on $E(\Phi_0)$, and similarly q_1 on components of $E(\Phi_1) - E(\Phi_0)$, this implies that q_∞ is one-to-one on $E(\Phi_1)$. Since later stages were constructed in the same manner, the same argument shows that for all m, q_∞ is one-to-one on $E(\Phi_{m+1}) - E(\Phi_m)$, and that this has image disjoint from that of $E(\Phi_m)$. Therefore q_∞ is one-to-one on the infinite union.

This leaves the additional points in the closure, $E - \bigcup_m E(\Phi_m)$, to be considered. As observed at the beginning of the proof, these are intersections of sets $T_{r(j),w(j)} \times J_{r,j}$. Such a limit point is really new (ie. not contained in some $E(\Phi_m)$) if the sequence r contains infinitely many "0" entries.

Suppose the jth entry is the first "0" in a sequence r. Then $T_{r(j),w(j)} \times J_{r,j}$ has image in the interior of the image of $H'_{(j-1):2,*} - H'_{j:2,*}$. The intersection therefore cannot be a point in $E(\Phi_0)$. Since there are infinitely many "0" entries, this argument can be repeated to show that the image of the intersection is not in the image of any $E(\Phi_m)$.

Finally we show that two different limit points must have different images. Again suppose the jth entry is the first "0" in r. If another sequence has j^{th} torus different from $T_{r(j),w(j)}$ then the images of the tori crossed with $J_{r,j}$ lie in distinct components of $H'_{(j-1):2,*} - H'_{j:2,*}$, so the image of the intersections cannot be the same point. Similarly if the solid tori in the two sequences differ after n "0" entries then they have distinct images under the singular parameterization constructed in step n, so the images of the intersections must be distinct. Again since there are infinitely many "0" entries then sequences of tori which ever differ must have distinct images. Therefore q is one-to-one on $E - \bigcup_m E(\Phi_m)$. This completes the proof of the proposition. ∎

Exercise Locate the step where the fact is used that C' is an infinite tower rather than just an infinite grope. This is the step which necessitates the development of towers of more than one story. □

4.4. The common quotient

In this section we construct maps $\alpha\colon D^2 \times D^2 \to Q$ and $\beta\colon C' \to Q$, which will be approximated by homeomorphisms in the next two sections. Composing these approximations will then give the homeomorphism required for 4.1.

A collection of subsets of a metric space is called *null* if for every $\epsilon > 0$ there are only finitely many of the subsets which have diameter larger than ϵ. Notice that only countably many of the subsets in a null collection can be bigger than a single point.

Lemma. *Suppose C' is a pinched regular neighborhood of a convergent tower, which can be parameterized as in 4.2. Then there are maps $\alpha\colon D^2 \times D^2 \to Q$ and $\beta\colon C' \to Q$ such that*

(1) *Q is a compact metric space,*
(2) *both α and β are onto, their collections of point inverses are null, points with nontrivial inverses are nowhere dense in Q and*
(3) *each nontrivial point inverse of α is either a regular neighborhood of a point (ie. a ball), or the union of an embedded 2-disk and a regular neighborhood of its boundary circle.*

Proof: Q will be defined as a quotient of $D^2 \times D^2$, with α the quotient map. Most of the proof is therefore involved with construction of subsets as described in (3), to be identified to points.

A D^2 union with a regular neighborhood of the boundary circle is $S^1 \times D^3 \cup D^2$, the union being over $S^1 \times \{y\}$ for some $y \in \partial D^3$:

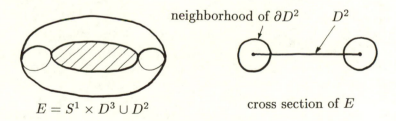

neighborhood of ∂D^2 D^2

$E = S^1 \times D^3 \cup D^2$ cross section of E

According to 4.3 the solid tori $T_{s,*}$ are regular neighborhoods of boundary circles of framed immersed 2-disks in $S^1 \times D^2$. Denote these 2-disks by $D_{s,*}$. Then the sets $T_{s,*} \times [.s1, .s2] \cup D_{s,*} \times \{.s11\}$ are the sorts of sets we want as inverse images of α. They cannot actually be inverse images because there are intersections and selfintersections. The first step is therefore to make them disjoint.

Choose continuous functions $e_{s,*}\colon D_{s,*} \to (.s0, .s22)$ satisfying the conditions

(i) $e_{s,m} = .s11$ on $D_{s,m} \cap T_{s,m}$, and is a constant non-endpoint element of the Cantor set on $D_{s,m} \cap T_{s,w}$, for $m \neq w$, and

(ii) the graphs of the $e_{s,m}$ are all disjoint.

These functions are constructed by induction on the length of s. Suppose they are chosen for sequences of length n, and consider a torus $T_{s,m}$ where $\ell s = n$. If t is a sequence with $dt = s$, we show how to choose e for the $D_{t,*}$ inside $T_{s,m}$.

By condition (i) the previous functions are constant on the intersection with $T_{s,m}$, and have values non-endpoints in the Cantor set. Therefore there are numbers $.s01 < a < .s1$ and $.s2 < b < .s22$ such that none of these previous functions take values in (a, b). By condition (1) of 4.3 the disks $D_{t,*}$ are contained in $T_{s,m}$. Therefore if the functions $e_{t,*}$ are constructed with values in (a, b) the graphs will be disjoint from graphs of the previous functions.

We are reduced to finding functions $e_{t,*}\colon D_{t,*} \to (a, b)$ which satisfy condition (i) above, and whose graphs are disjoint from each other.

According to 4.3 the disks $D_{t,*}$ intersect each other in disjoint arcs. Choose neighborhoods of segments of these arcs disjoint from the boundary, so that the complements of the neighborhoods are disjoint. Begin with $e = .s11$, and in these neighborhoods perturb e. On one disk perturb e into the future, the other into the past.

into past

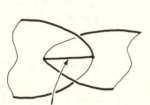 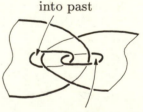

arc of intersection or selfintersection   into future

This is easily done to satisfy (i). The graphs are disjoint from each other, so as remarked above also satisfy (ii). Therefore the induction step in the construction of the $e_{t,*}$ is complete.

Now define Q as the quotient of $D^2 \times D^2$ obtained by identifying to distinct points the closure of the ball B, and each $T_{s,m} \times J_s \cup (\text{graph of } e_{s,m})$. The map α is the quotient map. The map β is the inverse of the partial parameterization, composed with α. This extends to a map of C' by mapping each component of the complement of the parameterized region to a point. This is possible since the frontier of the component is mapped to the frontier of B, or of some $T_{s,m} \times J_s$, which is mapped to a point in Q.

It is clear from the construction that Q satisfies all the conditions of the lemma, except possibly the nowhere density hypothesis and that it is a metric space. To prove Q is a metric space we will show it is Hausdorff, and appeal to the metrization theorem for compact Hausdorff spaces. It would be more in keeping with our explicit approach to directly construct a metric on Q, but this seems to be substantially harder.

Q is Hausdorff if distinct points have disjoint neighborhoods. Since distinct point inverses in D^4 have disjoint neighborhoods, it is sufficient to show that if U is a neighborhood of $\alpha^{-1}(q)$ in D^4, there is a neighborhood $V \subset U$ such that if $\alpha^{-1}(p)$ intersects V then it is contained in V. Let $\{E_i\}$ denote the point inverses which intersect U but are not contained in it (recall the nontrivial point inverses are countable). Define V to be $U - \bigcup_i E_i$. If V is open it is the required neighborhood. It can fail to be open only by containing a limit point of the E_i. Since the diameters of the point inverses go to 0, and since E_i intersects $D^4 - U$, distances of points in E_i to $D^4 - U$ goes to 0 as i goes to ∞. Therefore $\lim_{i \to \infty} E_i \subset D^4 - U$, and V is open.

To show nowhere density, suppose U is an open set in Q which contains a limit point of images of nontrivial point inverses. Since U contains one image, $\alpha^{-1}(U)$ is an open set containing one of the $T_{s,m} \times J_s \cup ($graph of $e_{s,m})$. Let $T_{ds,w}$ be the torus which contains $T_{s,m}$, then $\mathrm{int}(T_{ds,w} \times J_s) - \bigcup_* ((T_{s,*} \cup D_{s,*}) \times J_s) - ($graphs of $e_{r,*}; \ell(r) \le \ell(s))$ is an open set disjoint from all nontrivial point inverses, and with nonempty intersection with $\alpha^{-1}(U)$. Its image in U contains no nontrivial images, so these are not dense in U. ∎

4.5 Approximation of α

Proposition. *Let $\alpha: D^4 \to Q$ be as specified in 4.4. Then there is a homeomorphism $h: D^4 \to Q$ which agrees with α on ∂D^4.*

Proof: The basic idea is to find a map $f_\infty: D^4 \to D^4$ which has the same point inverses as α, and is the identity on the boundary. Then $h = \alpha(f_\infty^{-1})$ is a homeomorphism with the desired properties. The function f_∞ will be obtained as a limit of functions. We will reduce the construction of these functions to the careful construction of functions which have a single nontrivial point inverse. The key step is the following lemma;

Lemma. *Suppose $E \subset D^4$ is closed and starlike, or a flat 2-disk union a regular neighborhood of the boundary circle, $\{T_*\}$ is a null collection of closed sets disjoint from E, and $\delta > 0$. Then there is a map $k: D^4 \to D^4$*

which is the identity on the boundary, $k(E) = 0$, k is a homeomorphism on $D^4 - E$, and for each i either k moves points in T_i less than δ, or $k(T_i)$ has diameter less than δ.

Proof of the lemma: Suppose first that E is starlike. By *starlike* we mean that each ray from 0 intersects E in a closed interval containing 0. For a starlike set E there is a radius function $e: S^3 \to [0, 1]$ defined by $e(y) = \max\{t; ty \in E\}$. E is closed if and only if e is upper semicontinuous ($e^{-1}[0, s)$ is open, for all s). If e is continuous and nonvanishing then E is homeomorphic to D^4; the map $x \mapsto e(x/|x|)x$, for $x \neq 0$, is a homeomorphism.

Note that if E is closed and starlike, and U is a neighborhood of E, then there is a starlike V with continuous nonvanishing radius function such that $E \subset \text{int}V \subset U$. Such a V can be found by approximating the radius function e from above.

The map k in the lemma will be obtained by the following construction. Suppose W_i are starlike sets with continuous radius functions, $\text{int}W_i \supset W_{i+1}$, and $W_0 = D^4$. Suppose B_i are round balls centered at 0, with radii decreasing to 0, and $B_0 = D^4$. Then define $k: D^4 \to D^4$ by: for each ray from 0, map the intersection with $W_i - W_{i+1}$ affinely to the intersection with $B_i - B_{i+1}$, and map $\bigcap_i W_i$ to 0. These intersections are intervals of nonzero length, so k is one-to-one except on $\bigcap W_i$. Finally k is continuous because the radius functions of the W_i are.

Now we choose the B_i and W_i so that k will have the required properties with respect to the T_*.

Choose B_i to be the round ball of radius b_i centered at 0, where $b_0 = 1$, $b_i - b_{i+1} < \delta/8$, and the b_i decrease monotonically to 0.

Choose V_0 containing E in its interior, and such that if some T_j intersects V_0 then it has diameter less than $\delta/2$. Since the T_* are a null collection there is some neighborhood of E disjoint from the finitely many which have diameter at least $\delta/2$, and as remarked above this neighborhood contains a starlike subneighborhood with continuous radius function. Therefore such a V_0 exists. For $i > 0$ choose V_i inside the interior of V_{i-1}, containing E in its interior and within $(1/2)^i$ of E, and such that any T_j which intersects $D^4 - V_{i-1}$ is disjoint from V_i. Again the nullity of the T_* implies that there is a neighborhood of E with these properties, and we take a continuous starlike subneighborhood.

Define k as above taking the sets $V_i \cup B_i$ to the sets B_i. (In the notation above $V_i \cup B_i$ is W_i.)

The first observation is that if $x \in B_i - V_i$ for some i, then k moves x a distance less than $\delta/8$. Suppose i is the maximal such index for x, so x is not in B_{i+1}. Let Rx denote the ray through x, and consider the interval $Rx \cap (V_i \cup B_i - V_{i+1} \cup B_{i+1})$. The conditions on x imply that the upper endpoint of this interval lies on B_i, so it is a subinterval of $Rx \cap (B_i - B_{i+1})$. Since k maps the smaller interval to the larger, and since the diameter of the larger is less than $\delta/8$, x is moved by less than $\delta/8$.

Since all of the T_* of diameter at least $\delta/2$ are in $B_0 - V_0$, it follows that points in these are moved less than δ. To prove the lemma it is therefore sufficient to show that if T_i has diameter less than $\delta/2$ then $k(T_i)$ has diameter less than δ.

Suppose then that x and y are points in T_i which has diameter less than $\delta/2$. If both are contained in sets of the form $B_j - V_j$, then by the above they are moved less than $\delta/8$, so $d(k(x), k(y)) < d(k(x), x) + d(x, y) + d(y, k(y)) < \delta$. If one point, say x, is not in such a set then $T_i \subset V_j - V_{j+2}$ for some j (where possibly $j = -1$, in which case we set $V_j = D^4$). Note x is not in B_{j+2}. There are two cases to consider; either y is, or is not, in B_{j+2}.

Let x', y' denote the intersection of the rays Rx and Ry with the sphere ∂B_{j+2}. If y is not in B_{j+2} then since x' and y' are closer to the origin and lie on a round sphere, $d(x', y') < d(x, y)$. But $k(x)$ and $k(y)$ lie on intervals of length less than $2(\delta/8)$ with x', y' as endpoints (eg. $Rx \cap B_- B_{j+2}$). Therefore $d(k(x), x') < \delta/4$ and $d(k(y), y') < \delta/4$. Putting these together we get $d(k(x), k(y)) < \delta$.

If y is in B_{j+2} then $d(x, k(y)) \leq d(x, y) + d(y, k(y)) < 5\delta/8$. But $k(x)$ is closer to $k(y)$ than x is. This is because $k(x)$ is closer to the origin than x, but still outside the round ball B_{j+2}, and $k(y)$ is inside the ball.

This completes the proof of the lemma if E is starlike. If E is a disk union a regular neighborhood of its boundary then we apply the starlike case twice: the 2-disk is starlike, and E with the 2-disk identified to a point is ambiently homeomorphic to a starlike set. This latter fact can be seen using a mapping cylinder neighborhood of E in D^4. ∎

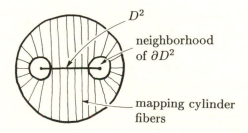

D^2

neighborhood
of ∂D^2

mapping cylinder
fibers

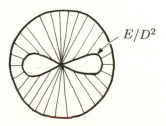

E/D^2

Exercise A subset $K \subset M^n$ is "starlike equivalent" if there is a homeomorphism of a neighborhood to D_n which carries K to a starlike set. It is "eventually starlike equivalent" if there is a finite closed filtration K_i so that $K_i/K_{i+1} \subset M/K_{i+1}$ is starlike equivalent, for each i. Suppose K has a neighborhood which is the mapping cylinder of a map $m\colon S^{n-1} \to K$. Give a criterion for K to be eventually starlike equivalent, in terms of the map m and homeomorphisms of K_i/K_{i+1} with starlike sets. Find such a criterion which applies to the non-starlike E in the lemma. □

Proof of the Proposition: To control the inverses of functions, which are only relations, we introduce some terminology. If R, S are relations from X to Y (ie. closed subsets of $X \times Y$), then $d(R, S) < \epsilon$ means that given y in one of $R(x)$ or $S(x)$ there is y' in the other so that $d(y, y') < \epsilon$. Here $R(x)$ is the intersection $R \cap \{x\} \times Y$, and d is the metric in Y.

Suppose there are functions $f_n\colon D^4 \to D^4$ which satisfy

(1) $d(f_n, f_{n+1}) < (1/2)^n$, $d(f_n^{-1}, f_{n+1}^{-1}) < (1/2)^n$, and f_n is the identity on ∂D^4,

(2) $f_n(\alpha^{-1}(q))$ has diameter less than $(1/2)^{n+1}$ for all $q \in Q$, and

(3) f_n has finitely many nontrivial point inverses, each equal to a point inverse of α, and all point inverses of α which have diameter at least $(1/2)^{n+1}$ occur this way.

Condition (3) implies that $\alpha(f_n^{-1})$ is a well-defined function, equal to α on ∂D^4. Q is a compact metric space so uniform continuity of α and condition (1) implies that $\{\alpha(f_n^{-1})\}$ is a Cauchy sequence, so converges to a function $h\colon D^4 \to Q$. Conditions (1) and (2) imply that the sequence of relations $\{f_n(\alpha^{-1})\}$ is a Cauchy sequence of relations, and converges to a function. This function is an inverse for h, so we conclude that h is a homeomorphism. The proposition will therefore be proved if we can find such functions f_n.

The f_n are constructed by induction. If f_{n-1} is given then we define $f_n = g f_{n-1}$, where g satisfies:

(a) g is the identity except on finitely many sets B_i disjoint from ∂D^4 and each other, and such that the diameters of B_i and $f_{n-1}^{-1}(B_i)$ are less than $(1/2)^n$,

(b) $g(f_{n-1}\alpha^{-1}(q))$ has diameter less than $(1/2)^{n+1}$ for all $q \in Q$, and

(c) there is exactly one nontrivial g point inverse in each B_i, each one equal to some $f_{n-1}\alpha^{-1}(q)$, and if the diameter of $\alpha^{-1}(q)$ is at least $(1/2)^{n+1}$ and $f_{n-1}\alpha^{-1}(q)$ is not already a point, then it occurs as one of the g point inverses.

Then it is easily seen that f_n satisfies properties (1)–(3).

Let E_i denote the nontrivial sets $f_{n-1}(\alpha^{-1}(q))$ such that either it or

$\alpha^{-1}(q)$ has diameter at least $(1/2)^{n+1}$. By condition (2) E_i has diameter less than $(1/2)^n$, so it has a neighborhood U_i which also has diameter less than $(1/2)^n$. By condition (3) $f_{n-1}^{-1}(E_i)$ has diameter less than $(1/2)^n$, so we can choose U_i so that $f_{n-1}^{-1}(U_i)$ also has this property. There are finitely many of the E_i, and they are closed and disjoint. There are also finitely many points whose f_{n-1} inverses are nontrivial, and these are disjoint from the E_i. Therefore we can choose the U_i to be mutually disjoint, and such that f_{n-1} is a homeomorphism on the preimages. Finally since the sets $f_{n-1}(\alpha^{-1}(q))$ form a null collection, we can choose the U_i so that each set $f_{n-1}(\alpha^{-1}(q))$ intersects at most one of the U_i, and the ones which do intersect have diameter less than $(1/2)^{n+2}$.

According to 4.4 the nontrivial $\alpha^{-1}(q)$ are either regular neighborhoods of points, or embedded 2-disks union a regular neighborhood of the boundary circle. Since f_{n-1} is a homeomorphism on a neighborhood of the ones not taken to points, the same is true of the E_i. Therefore there is a closed neighborhood B_i of E_i inside U_i with a homeomorphism $e: B_i \to D^4$, such that $e(E_i)$ is a standard model $E \subset D^4$. (B_i is just a slightly larger regular neighborhood of the point or embedded 2-disk.)

Choose $\delta > 0$ such that points within δ in D^4 have preimages within $(1/2)^{n+3}$. Define a null collection in D^4 by $T_* = e(f_{n-1}(\alpha^{-1}(q)))$, and let $k: D^4 \to D^4$ be a map as in the lemma which kills E and satisfies the conclusions about T_* and δ. Then g_i, defined to be the identity off B_i and $e^{-1}ke$ on B_i, kills E_i and does not enlarge any $e(f_{n-1}(\alpha^{-1}(q))$ to have diameter greater than $(1/2)^{n+2}$.

Finally if g is defined to be the composition of these g_i, it satisfies conditions (a)–(c). This completes the proof of 4.5. ∎

Exercise (1) Modify the proof to apply to manifolds of arbitrary dimension. (2) Show that if $\alpha: M \to Q$ is a map whose point inverses are null and eventually starlike equivalent, then α can be replaced by a homeomorphism. ☐

4.6. Approximation of β

Since Q is homeomorphic to D^4 (by 4.5), β is a map $C' \to D^4$ which restricts to a homeomorphism $\partial C' \to S^3$. In this section we see that there is a homeomorphism $h: C' \to D^4$ which agrees with β on C'. The situation is quite different from the previous section in that there we understood the point inverses very well, and needed information about the image space. Here we know the image, but the point inverses are mysterious.

The general case will be reduced to a $D^4 \to D^4$ situation:

Proposition. *Suppose* $f: D^4 \to D^4$ *is onto, has null point inverses, points with nontrivial inverses are nowhere dense, and* $E \supset S^3$ *is a closed set such that* $f: f^{-1}(E) \to E$ *is a homeomorphism. Then there is a homeomorphism* $h: D^4 \to D^4$ *which agrees with* f *on* E.

Before we prove this we indicate how it can be used to find an approximation for β. Define $f: C' \cup (\partial C' \times I) \to D^4 \cup (S^3 \times I) \cong D^4$ to be β on C' and $\partial\beta \times 1$ on the product. Since β is a homeomorphism on $\partial C'$, f is a homeomorphism over the set $\partial E = C' \times I$. If there is a homeomorphism $h: C' \cup (\partial C' \times I) \to D^4$ which agrees with f on $f^{-1}(E)$, then the restriction to C' gives a homeomorphism $C' \to D^4$ which restricts to $\partial\beta$ on the boundary. Therefore it is sufficient to show that $C' \cup (\partial C' \times I)$ is homeomorphic to D^4, so that we can obtain such an h from the proposition.

According to 4.3 C' is obtained from D^4 by dividing out arcs in a wild collar on S^3. Let $c: S^3 \times [0, 1] \to D^4$ be the collar, with $c(S^3 \times \{1\}) = S^3$, and let $K \subset S^3$ be the closed set so that C' is obtained by identifying to points the arcs of the form $c((k) \times [0, 1])$ for $k \in K$. Extend c to the collar $c: S^3 \times [0, 2] \to D^4 \cup (S^3 \times [1, 2])$, then $C' \cup (\partial C' \times I)$ is obtained by dividing out arcs $c(\{k\} \times [0, 1])$. As in the proof of 4.5, we get the required homeomorphism by constructing a map of $D^4 \cup (S^3 \times [1, 2])$ to itself with these arcs as nontrivial point inverses.

Let $j: S^3 \to [0, 1]$ have $j^{-1}(0) = K$ (recall K is a closed set). Define a map of $S^3 \times [0, 2]$ to itself by taking the interval $\{s\} \times [0, 1]$ linearly to $\{s\} \times [0, j(s)]$, and $\{s\} \times [1, 2]$ linearly to $\{s\} \times [j(s), 2]$. The desired map on $D^4 \cup (S^3 \times [1, 2])$ is this map on $c(S^3 \times [0, 1]) \cup (S^3 \times [1, 2])$, and the identity on the complement.

This completes the demonstration that the proposition implies that β can be replaced by a homeomorphism. ∎

Proof of the proposition: We begin with some terminology;

Two subsets of X are *well separated* if the closure of each is disjoint from the other. Note however that if we take the closure of both then intersections may occur.

The *singular image* Σf of a map f is the set of y such that $f^{-1}(y)$ is different from a point.

A diagram

$$D^4 \xrightarrow{\ e\ } W \xleftarrow{\ f\ } D^4$$
$$\downarrow g$$
$$D^4$$

together with a closed subset $E \subset D^4$ containing S^3 is called *admissible* if the point inverses of ge and gf are null collections, the singular images Σg, $g(\Sigma e)$ and $g(\Sigma f)$ are nowhere dense, disjoint from E, and well separated from one another, and on S^3 both ge and gf are the identity.

The plan is to perform modifications which make e and f converge to homeomorphisms. The role of g is to measure the progress toward this goal.

The point of the separation requirement is that if $w \in W$ is a singular point of e, then g and f^{-1} are homeomorphisms near w. The key step in the proposition is a lemma which eliminates some point inverses of e, at the expense of complicating f.

Lemma. *Suppose (W, e, f, g, E) form an admissible diagram, $B \subset D^4$ is a ball disjoint from E, Σg and $g(\Sigma f)$, and B_0 is a ball inside B such that $e: (ge)^{-1}(\partial B_0) \to g^{-1}(\partial B_0)$ is a homeomorphism. Then there is an admissible diagram (W', e', f', g', E) with an identification of $g^{-1}(D^4 - \mathrm{int} B_0) \subset W$ with $(g')^{-1}(D^4 - \mathrm{int} B_0) \subset W'$ such that $e = e'$ on $(ge)^{-1}(D^4 - \mathrm{int} B_0)$, $g = g'$ on $(g')^{-1}(D^4 - \mathrm{int} B_0)$, $f = f'$ on $(gf)^{-1}(D^4 - \mathrm{int} B)$, and e' is a homeomorphism on $(ge')^{-1}(B_0)$.*

Proof of the lemma: Since we can change W it is easy to change e to be a homeomorphism on the inverse of B_0 by redefining it to be the identity there. Precisely, we let $W' = [W - g^{-1}(\mathrm{int} B_0)] \cup [(ge)^{-1} B_0]$, with $y \in (ge)^{-1} \partial B_0$ identified with $e(y) \in g^{-1} \partial B_0$. Then e factors as

$$D^4 \xrightarrow{\ e \cup (\mathrm{id})\ } W' \xrightarrow{\ (\mathrm{id}) \cup e\ } W.$$

Define e' to be $e \cup (\ \mathrm{id})$, and g' to be $g \cup (ge)$, then we get a diagram

$$
\begin{array}{ccccc}
D^4 & \xrightarrow{\ e'\ } & W' & & \\
\downarrow{\scriptstyle =} & & \downarrow{\scriptstyle (\mathrm{id}) \cup e} & & \\
D^4 & \xrightarrow{\ e\ } & W & \xleftarrow{\ f\ } & D^4 \\
& & \downarrow{\scriptstyle g} & & \\
& & D^4 & &
\end{array}
$$

An appropriate $f': D^4 \to W'$ will be constructed by composing f with a map $W \to W'$.

Suppose $w: B \to D^4$ be a homeomorphism which is the identity on B_0. Define $h: W \to W$ to be the identity on $g^{-1}(D^4 - \mathrm{int} B)$, and $h = g^{-1} w^{-1} gewg$ on $g^{-1}(B)$. (To check that these match up on overlaps recall that ge is the identity on S^3 and that e is a homeomorphism on

$(ge)^{-1}(\partial B_0)$.) In the part which maps into $g^{-1}(B_0)$ the w^{-1} in the $g^{-1}w^{-1}g$ factor is the identity, so this piece cancels out. This leaves ewg, which has the lift wg into $(ge)^{-1}B_0$. This defines a lift $h': W \to W'$, and we define $f' = h'f$.

We must check (or arrange) the nullity and well separation hypotheses for f'. Note that $g'e' = ge$, and $\Sigma g' \cup g'(\Sigma e') = \Sigma g'e' = \Sigma ge$, so for the separation hypothesis we want $g'(\Sigma f')$ well separated from Σge. On the domain mapping into $(g')^{-1}(D^4 - \mathrm{int}B)$, $f' = f$ and the original hypotheses apply. On the domain mapping into $(g')^{-1}(\mathrm{int}B_0)$, $f' = wgf$, which is a homeomorphism there.

Consider therefore the region mapping into $g^{-1}B - g^{-1}(\mathrm{int}B_0)$, where $f' = g^{-1}w^{-1}gewgf$. On this domain wgf is a homeomorphism, as is $g: g^{-1}B \to B$, so it is sufficient to consider $w^{-1}ge: D^4 - B_0 \to B - B_0$. This has the same point inverses as ge, so they are null. Since w is a homeomorphism the singular image is $w^{-1}\Sigma ge = w^{-1}g\Sigma e$, and we want this well separated from $(g\Sigma e) \cap (B - B_0)$. We claim that w can be chosen so this holds.

The set of possible w (homeomorphisms $B \to D^4$ which are the identity on B_0) is a complete metric space. A metric is given by: the distance from V to w is $\max(d(w,v), d(w^{-1}, v^{-1}))$. Since $g\Sigma e$ is nowhere dense, if $p \in (B - B_0)$ then the set of w such that $w(p) \notin \mathrm{cl}(g\Sigma e)$ is open and dense in the space. Similarly given q, the w with $w^{-1}(q) \notin \mathrm{cl}(g\Sigma e)$ is open and dense. Since $g\Sigma e$ is countable, the set of w with $g\Sigma e \cap \mathrm{cl}(w^{-1}g\Sigma e) = \mathrm{cl}(g\Sigma e) \cap w^{-1}g\Sigma e = \phi$ is a countable intersection of open dense sets, hence dense by the Baire category theorem. Therefore w can be chosen so that f' satisfies the well separation hypothesis. This completes the proof of the lemma. ∎

Proof of the Proposition, continued: We set up an admissible configuration as the starting point for an inductive construction. Let $W_0 = D^4$, $f_0 = f$ (the f given in the proposition), $e_0 = c(\partial f)$, $g_0 = c(\partial f)^{-1}$, and $E_0 = c(\partial f)^{-1}f(E)$. Here $c(\partial f)$ is the cone on $f: S \to S$, which is a homeomorphism, and $c(\partial f)^{-1}$ is the inverse.

Suppose there are admissible configurations $(W_i, e_i, f_i, g_i, E_0)$ which satisfy

(1) $d(f_i^{-1}e_i, f_{i+1}^{-1}e_{i+1}) < (1/2)^i$, $d(e_i^{-1}f_i, e_{i+1}^{-1}f_{i+1}) < (1/2)^i$,

(2) $\mathrm{diam}(e_i^{-1}(w)) < (1/2)^i$, $\mathrm{diam}(f_i^{-1}(w)) < (1/2)^i$ for $w \in W_i$, and

(3) $(g_if_i)^{-1}(E_0) = (g_0f_0)^{-1}(E_0)$, and $e_i^{-1}f_i = e_0^{-1}f_0$ on this set.

Note that $f^{-1}e$, $e^{-1}f$ are relations. The distance between relations used in (1) is defined in 4.5. Condition (1) implies that the relations $e_i^{-1}f_i$ converge to a relation h'. Condition (2) implies that h' is a homeomorphism. Finally (3) implies that $h = e_0^{-1}f = c(\partial f)^{-1}f$ on E ($= (g_0f_0)^{-1}(E_0)$).

Therefore if we define $h = c(\partial f)h'$ then h satisfies the conclusions of the proposition.

Suppose that configurations satisfying the conditions are defined for $i \leq n$. We construct the $n+1$ configuration by applying the lemma.

Let x_i, y_i be the points not in Σg_n such that one of $(g_n f_n)^{-1}(x_i)$ or $(g_n e_n)^{-1}(y_i)$ has diameter at least $(1/2)^{n+1}$. By the separation hypothesis all point inverses of e_n or f_n with diameter at least $(1/2)^{n+1}$ occur as the inverse image of $(g_n)^{-1}(x_i)$ or $(g_n)^{-1}(y_i)$. The nullity hypothesis implies that there are only finitely many of them.

Suppose $\delta > 0$, and let Bx_i, By_i denote the balls of radius δ. There is a $\delta > 0$ small enough so that these balls are disjoint from each other, E_0, Σg_n, and intersect only one of $g \Sigma e_n$ and $g \Sigma f_n$. Since $(g_n e_n)^{-1}(x_i)$ has diameter less than $(1/2)^n$, we can choose δ small enough so that $(g_n e_n)^{-1}(Bx_i)$ has diameter less than $(1/2)^n$. Since $(g_n e_n)^{-1}(y_i)$ is a point, we can arrange $(g_n e_n)^{-1}(By_i)$ to have diameter less than $(1/2)^{n+1}$. Similarly we can arrange $(g_n f_n)^{-1}(By_i)$ and $(g_n f_n)^{-1}(Bx_i)$ to have diameters less than $(1/2)^n$ and $(1/2)^{n+1}$ respectively.

Finally choose $\delta > \delta' > 0$ so that the boundaries of the balls of radius δ' are disjoint from $g_n \Sigma e_n$ and $g_n \Sigma f_n$. This is possible since these sets are countable, and there are uncountably many δ' in $(0, \delta)$.

The balls of radius δ and δ' about x_i satisfy the hypotheses of the lemma. Therefore there is an admissible configuration (W', e', f', g', E_0) in which e' is a homeomorphism where e_n had large point inverses. Interchange e' and f' to get another admissible configuration. Now the balls of radius δ and δ' about y_i satisfy the hypotheses of the lemma for (W', f', e', g', E_0). Apply the lemma to give a configuration in which f' has been changed to be a homeomorphism where it had large point inverses. Switch the e and f back and call the result $(W_{n+1}, e_{n+1}, f_{n+1}, g_{n+1}, E_0)$. The restrictions on the size of inverses of the balls (where the maps have been changed) easily imply that this configuration satisfies the conditions (1)-(3).

By induction an entire sequence of admissible configurations exists. As indicated above this implies the truth of the proposition, and completes the approximation of β. ∎

Exercise (1) Show that the h in the proposition can be chosen to approximate f within any given $\epsilon > 0$. (2) Modify the proof to apply to disks of arbitrary dimension. □

4.7. References

The general outline of section 4 is that of the original proof in Freedman [2]. The singular parameterizations of 4.2 are less singular than the original in that they are homeomorphisms on the boundary. This improvement is a result of the use of towers of gropes rather than towers consisting only of disks. In the original the partial parameterizations analogous to 4.3 are still singular, which forces the analogs of 4.4–4.6 to be considerably more complicated. The decomposition parts of Freedman [2] are done in the context of the Bing Shrinking Theorem, and many abstract results. The simplifications here have permitted a more direct and concrete approach.

Ancel and Starbird [1] have characterized the number of grope stages needed in a tower for the parameterization to be nonsingular on the boundary.

The singular parameterizations of 4.2 identify the sets C' as *crumpled cubes* in the terminology of Davermann [1]. See Cannon, Bryant, and Lacher [1] for crumpled cubes arising as pinched neighborhoods of convergent gropes in higher dimensional manifolds. Cannon [1] has some related expository material. The lemma on Bing doubles comes from one of the first papers on what has come to be known as Bing topology, Bing [1].

The material on "starlike" decompositions in 4.5 is decended from Bing [2] and Bean [1]. The idea behind the disk-to-disk proposition in 4.6 comes from Freedman [2, §8], see also Ancel [1] for a detailed treatment. This result is a very special case of the cell-like approximation theorem of Quinn [4, 2.6.2], but in dimension 4 the general theorem depends on the special case.

We note the use of towers and much of the decomposition material might conceivably be avoided. Convergent infinite gropes (see the exercise in 3.8) are more easily constructed, and have frontiers which are properly parameterized (see the last exercise in 4.2). These objects have "small transverse spheres" near the limit points, which implies they are 1-LC embedded. According to the flattening theorem 9.3B this implies the frontier is collared in the complement. A correctly framed embedded disk is easily found inside such a collar, so this implies the embedding theorem.

The difficulty with this sketch is that the proof given for the flattening theorem depends heavily on the embedding theorem. There are weaker flattening theorems independent of the embedding theorem, eg. Kirby [3], but they do not produce the required collars. To get a proof from this approach, the problem is therefore to find a proof of a sufficiently strong flattening theorem which does not use the embedding theorem.

This simplification would only be a pedagogical improvement: the construction of convergent infinite gropes still requires the poly-(finite or cyclic) fundamental group hypothesis.

CHAPTER 5

The Embedding Theorems

The keys to the study of 4-manifolds are embedding theorems for 2-disks, as explained in the introduction. In this chapter precise statements of the embedding results are given, and proofs assembled from the rest of Part I. In all cases the goal is to find a collection of disjointly embedded disks. Some statements can be extended to other surfaces, but because of their crucial importance we focus on disks.

The technically important results are the poly-(finite or cyclic) embeddings of section 5.1, and the controlled embeddings of 5.4. The results of 5.2 and 5.3, in which hypotheses are imposed on images in the fundamental groups, have had some application. However their main interest is that they seem to represent the limits of what can be achieved with these techniques.

The proofs depend on the immersion lemma 1.2. The results are therefore unconditionally valid for embeddings in PL or smooth manifolds. For topological manifolds this lemma is proved in Chapter 9, using the embedding theorems for smooth or PL manifolds. The topological case is therefore logically dependent on this material (and should not be used in the proofs of it).

5.1. Poly-(finite or cyclic) fundamental groups

We recall that a group is poly-(finite or cyclic) if there is a finite ascending sequence of subgroups, each normal in the next, and with adjacent quotients either finite or cyclic.

5.1A Theorem. *Suppose $A \to M^4$ is an immersion of a union of disks, with algebraically transverse spheres whose algebraic intersections and selfintersection numbers are 0 in $\mathbf{Z}[\pi_1 M]$. If $\pi_1 M$ is poly-(finite or cyclic) then there is a topologically embedded union of disks with the same framed boundary as A, and with transverse spheres.*

We spell out the hypotheses of 5.1: Index the components of A as A_i, $i = 1 \ldots n$. The algebraic transverse sphere hypothesis is equivalent (by 1.9) to the existence of elements $\alpha_j \in \pi_2 M$ such that $\lambda(A_i, \alpha_j) = 0$

if $i \neq j$, and $= 1$ if $i = j$, $\lambda(\alpha_i, \alpha_j) = 0$, and $\mu(\alpha_i) = 0$. These are purely algebraic-topological conditions, and amenable to manipulation. For example this statement for many disks can be deduced from the statement for a single disk, as given in the introduction. We have stated it here for many disks because it is used in that form, and for uniformity with results which cannot be reduced.

A useful modification of the $\pi_1 M$ hypothesis is described in the exercise at the end of the proof.

Proof: We work backwards through the logical structure of the proof. Theorem 4.1 gives correctly framed topologically embedded disks in a pinched regular neighborhood of a convergent tower. Proposition 3.8 gives convergent towers from properly immersed 1-story capped towers. Lemma 3.3 gives a 1-story capped tower from a π_1-null 3-stage capped grope with transverse spheres. Since the fundamental group of M is poly-(finite or cyclic), proposition 2.9 shows that a 3-stage capped grope can be changed to be π_1-null. It remains to show that under the hypotheses of the theorem there is a properly immersed (union of disks)-like capped grope with 3 stages, transverse spheres, and the same framed boundary as A. This is assigned as an exercise in 2.5, but we sketch the construction.

At each intersection point of A add a copy of the algebraically transverse sphere to one sheet. The resulting disks have trivial intersection and selfintersection numbers, the same framed boundary, and the spheres are still algebraically transverse to them. Construct immersed Whitney disks for the extra intersections between the disks and spheres, and use the immersed Whitney move to remove these intersections. The resulting spheres are (geometrically) transverse spheres for the disks, so will be denoted A^t. Next choose Whitney disks for all intersections among the A, and use these to do 1-surgeries to convert A into a capped surface, as in 2.1. Use sums with A^t to get caps disjoint from the body surface.

Essentially we repeat the argument for the caps, and convert them into capped surfaces. The first step is to get the spheres A^t disjoint from the caps. At each intersection of A^t with a cap, push down to the first stage and add parallel copies of A^t. Each intersection leads to the addition of two copies of a sphere (see 2.5) so all intersections introduced this way occur in pairs with a Whitney disk. The resulting transverse spheres thus still satisfy the selfintersection conditions and additionally have algebraically trivial intersection with the caps. Next change the caps by pushing down intersections and adding copies of these new A^t to get caps with trivial algebraic selfintersections. Use the Whitney move to separate the caps from the transverse spheres. Choose Whitney disks for selfintersections of the caps, and do 1-surgeries as in 2.1 to convert the caps to capped surfaces. Finally move the new caps off the surfaces by adding copies of A^t.

At this point we have a properly immersed 2-stage capped grope, with transverse spheres which may intersect the caps. Repeat the argument above to get height 3 with disjoint transverse spheres, which completes the proof. ∎

Exercise   An *enlargement* of a 4-manifold M is an inclusion as an open set in another 4-manifold, $M' \supset M$. Note $\partial M = M \cap \partial M'$. Show that if the data of 5.1A is given in M, and M' is an enlargement such that the image of $\pi_1 M \to \pi_1 M'$ is poly-(finite or cyclic) then the conclusion of 5.1A holds in M'. ☐

In the theorem, all the algebraic intersection hypotheses apply to the transverse spheres. Some of these can be shifted to A;

5.1B Corollary. *Suppose $A \to M^4$ is an immersion of a union of disks, whose algebraic intersections and selfintersection numbers are 0, and with algebraically transverse spheres. If $\pi_1 M$ is poly-(finite or cyclic) then A is regularly homotopic rel boundary to a topologically embedded union of disks with transverse spheres.*

Proof: Use immersed Whitney moves to remove extra intersections of A and the algebraically transverse spheres. This changes A by regular homotopy, and the result has transverse spheres A^t. The algebraic intersection hypotheses imply that there are immersed Whitney disks for all intersections of A, and the interiors of these can be made disjoint from A by sums with the A^t. For each such Whitney disk consider the linking torus of an intersection point as a transverse capped surface (see 2.1). Use the A^t to get caps for these disjoint from A, and contract to get algebraically transverse spheres for the Whitney disks, with algebraically trivial intersections and selfintersections. Denote by M_0 the complement of an open regular neighborhood of A in M. Then the Whitney disks and the algebraically transverse spheres give an immersion in M_0 satisfying the hypotheses of 5.1. The conclusion of 5.1 is that there are topologically embedded Whitney disks for the A intersections. Use these for Whitney moves to produce a regular homotopy of A to an embedding. ∎

5.2. π_1-null conditions

We expect the statements above to be false without the restriction on fundamental group, although this remains unsettled at the time of writing. In this section we give results in which the fundamental group is arbitrary, but conditions are placed on the image in the fundamental group of various parts of the data.

5.2A Theorem. *Suppose $A \to M^4$ is an immersion of a union of disks, with π_1-null transverse capped surfaces. Then there is a topologically embedded union of disks with the same framed boundary as A, and with transverse spheres.*

Recall (see 3.3) that transverse capped surfaces have body surfaces which are disjointly embedded and intersect A in single points. Caps for these surfaces must have interiors disjoint from the bodies, but are allowed to intersect A. π_1-null means loops in the image are contractible in M. This is a very strong assumption.

Proof: Proposition 3.3 gives a properly immersed capped tower with transverse spheres, and as in the proof of 5.1 this implies there are framed embedded disks. ∎

As in the corollary to the poly-(finite or cyclic) theorem, we can shift some of the hypotheses from the transverse object to A.

5.2B Corollary. *Suppose $A \to M^4$ is an immersion of a union of disks, whose algebraic intersections and selfintersection numbers are 0, and A^t are transverse spheres such that $A \cup A^t$ is π_1-null. Then A is regularly homotopic rel boundary to a topologically embedded union of disks with transverse spheres.*

Proof: Choose Whitney disks for the A intersections, and make the interiors disjoint from A by adding copies of the transverse spheres. These have transverse capped surfaces with bodies linking tori of intersection points, and caps normal disks to A, as in 3.1. Add copies of the transverse spheres A^t to get caps for these surfaces disjoint from A. Denote by M_0 the complement of an open regular neighborhood of A in M, then the Whitney disks and the transverse capped surfaces satisfy the hypotheses of 5.2. The π_1-nullity hypothesis applies since the transverse capped surfaces are contained in a neighborhood of $A \cup A^t$, which is π_1-null, and $M_0 \to M$ is an isomorphism on π_1.

Applying theorem 5.2 yields embedded Whitney disks for A, with transverse spheres disjoint from A. Use these transverse spheres to alter A^t to be disjoint from the Whitney disks. Whitney moves across the disks give a regular homotopy of A to an embedding, with transverse spheres. ∎

5.3. Embedding up to s-cobordism

The next result weakens the hypotheses of Corollary 5.2B a little further, at the expense of weakening the conclusion.

An *s-cobordism* (rel boundary) is a compact manifold N^5 with boundary in three pieces $\partial N = \partial_0 N \cup \partial_1 N \cup P \times I$ such that $\partial_i N \cap P \times I = P \times \{i\}$, and each inclusion $\partial_i N \to N$ is a simple homotopy equivalence. If M^4 is not compact then an s-cobordism of M with *compact support* is an s-cobordism $(N; \partial_0 N, \partial_1 N, P \times I)$ where $\partial_0 N$ a compact submanifold of M, union with $(M - \partial_0 N) \times I$. An *s-cobordism of an immersion* $A \to \partial_0 N$ is an s-cobordism N and an immersion $A \times I \to N$ which is a product on $\partial A \times I \to P \times I$.

A major potential application of disk embedding theorems is to show that s-cobordisms are products. Allowing the manifold to change by s-cobordism rules out this use, but there are other important potential uses ("surgery").

Theorem (Embedding up to s-cobordism). *Suppose $A \to M^4$ is an immersion of a finite union of disks, whose algebraic intersections and selfintersection numbers are 0, and A^t are algebraically transverse spheres such that $A \cup A^t$ is π_1-null. Then there is a compactly supported s-cobordism rel boundary of A to a topologically embedded union of disks with transverse spheres.*

The advantage of this statement over corollary 5.2B is that A^t need only be algebraically rather than geometrically transverse to A. The π_1-null hypothesis is too restrictive for the theorem as stated to go very far, though it will be used in section 12.4 to show a particular case of Poincaré transversality is equivalent to "surgery."

The proof of 5.3 is postponed to the next chapter, since it is a substantial diversion from the main line of argument.

5.4. Controlled embeddings

Suppose X is a metric space, and a function $\epsilon: X \to (0, \infty)$ is given. We want to show that for certain 4-manifolds $M \to X$ there are embedded disks in M whose images in X have diameter less than ϵ. By "diameter less than ϵ" for a subset $Y \subset X$ we mean that Y has diameter less than $\epsilon(y)$ for all $y \in Y$.

Fundamental groups play an important role, and to control this a reference space $p: E \to X$ is introduced. The map p is said to have "local fundamental groups in a certain class" if for every point and neighborhood $x \in U \subset X$ there is a neighborhood $x \in V \subset U$ such that the image of $\pi_1 p^{-1}(V) \to \pi_1 p^{-1}(U)$ is in the class. (If the inverse images are not connected we require this for each component.)

This control in E is related to a particular manifold M by specifying a map $M \to E$ which is "$(\delta, 1)$-connected" over X. This means that given a relative 2-complex (K, L) and a commutative diagram

then there is a map $K \to M$ which extends the given map on L and such that the composition $K \to M \to E$ is within δ of the original.

We also have to specify some algebraic intersection information. We will not define "δ intersection numbers" in general, but will say that two surfaces have "δ-algebraically trivial" intersections if all intersection points can be arranged in pairs with immersed Whitney disks each of diameter less than δ. Similarly "δ-algebraically transverse spheres" for a surface is a collection of framed immersed spheres with a distinguished intersection point in each component, and immersed Whitney disks of diameter less than δ for all other intersections.

Theorem (Controlled embeddings). *Suppose X is a locally compact metric space, $p: E \to X$ has poly-(finite or cyclic) local fundamental groups, and $\epsilon: X \to (0, \infty)$ is given. Then there is $\delta: X \to (0, \infty)$ such that if $M^4 \to E$ is $(\delta, 1)$-connected over X, $A \to M^4$ is a locally finite immersion of a union of disks, with diameters less than δ and δ-algebraically transverse spheres with δ-algebraically trivial intersections, then there is a topologically embedded union of disks with the same framed boundary as A, and with transverse spheres, of diameter less than ϵ.*

Proof: As in 5.1 we work back from the conclusion to the initial hypotheses. If we can find 1-stage capped towers with height 4, disjoint components of diameter less than ϵ and the right framed boundaries, then there are disjointly embedded disks of diameter less than ϵ. This is because 4.1 and 3.8 provide embedded disks inside regular neighborhoods of the components of the capped tower. Proposition 3.6 shows that there is $\gamma_1: X \to (0, \infty)$ so that the existence of γ-π_1-null capped gropes of height 4 with transverse spheres implies the existence of such towers.

Now the local fundamental group hypothesis is used. There is a $\gamma_2: X \to (0, \infty)$ so that if $Y \subset X$ has diameter less than γ_2 then the image of $\pi_1 p^{-1}(Y)$ in the fundamental group of the inverse of some set of diameter less than γ_1 is poly-(finite or cyclic). If we can find capped gropes with disjoint components, of diameter less than γ_2, then we can apply the fundamental group reduction proposition 2.9 in a small neighborhood of each component to get a γ-π_1-null capped grope. Further, this can be done to preserve small transverse spheres.

The controlled separation lemma 2.10 implies that there is γ_3 so that if we can find a properly immersed capped grope of height 3 with component images of diameter less than γ_3, then we can find one with disjoint components of diameter less than γ_2. Again this can be done to preserve transverse spheres.

We are reduced to showing that there is $\delta\colon X \to (0, \infty)$ so that data as given in the theorem yield a 3-stage properly immersed grope with component images of diameter less than γ_3. This is not difficult, and is assigned as an exercise in 2.10. This proof is the same as the corresponding construction in the proof of 5.1, noting that there are a limited number of steps—we count about 15—and each increases diameter in a predictable way. See section 2.10 for examples of such estimates. ∎

5.5. References

The embedding theorem for disks in a simply connected manifold appears in Freedman [2], and the poly-(finite or cyclic) case is sketched in Freedman [4]. The π_1-null theorem has several antecedents, eg. Freedman [4], theorem 5. The statement given—which we suspect to be the best possible without fundamental group restrictions—is new here.

Embedding up to s-cobordism is also considered here for the first time, but the result is essentially equivalent to theorem 1 of Freedman [6]. The controlled embedding theorem comes from Quinn [3].

CHAPTER 6

Embedding up to s-Cobordism

In this chapter we prove theorem 5.3, which asserts that a π_1-null collection of immersed disks and algebraically transverse spheres, with algebraically trivial intersections, is s-cobordant to an embedding. Some comments are given in 6.4 to explain how the proof breaks down without the π_1-null hypothesis.

6.1 Embeddings of transverse pairs

The model for a *transverse pair* is two copies of $S^2 \times D^2$ plumbed together at one point. Note each sphere is a transverse sphere for the other. This model is also a neighborhood of the pair of spheres $S^2 \times \{pt\} \cup \{pt\} \times S^2 \subset S^2 \times S^2$, which demonstrates that the boundary of the regular neighborhood of a transverse pair is S^3.

A "union of transverse pairs" is a disjoint union of copies of this model, and an immersed union of transverse pairs is obtained by introducing further plumbings. We say that an immersed union of transverse pairs has algebraically trivial intersections if the further plumbings can be arranged in pairs with Whitney disks. Note that this can also be described as two collections of immersed spheres with particular intersection and selfintersection numbers. The "transverse pair" terminology is introduced to avoid having to repeat the intersection number description.

Theorem. *A π_1-null immersion of a union of transverse pairs, with algebraically trivial intersections, is s-cobordant to an embedding.*

The proof is given after some general material on s-cobordisms.

6.2 Construction of s-cobordisms

Suppose $(N^5; M_0, M_1, P \times I)$ is an s-cobordism. There is a handlebody structure for the pair (N, M_0) with 2- and 3-handles, which can be indexed so that the boundary homomorphism in the cellular chain complex $C_3^c(N, M_0) \to C_2^c(N, M_0)$ (with $\mathbf{Z}[\pi_1 N]$ coefficients) is the identity matrix. This is part of the proof of the high-dimensional theorem, and this much of it works in this dimension. Here we use the (much easier) converse; a manifold with such a handle decomposition is an s-cobordism.

It is easy to show that the fundamental groups and $\mathbf{Z}[\pi_1 N]$ homology groups of M_i and N are the same, so the inclusions are homotopy equivalences. The Whitehead torsion of these inclusions is represented by the boundary homomorphism in the relative chain complex, so is trivial.

We set up some notation for handlebody structures on s-cobordisms. Begin with a 4-manifold M_0, with boundary P. Attach 2-handles $D^2 \times D^3$ on embeddings $S^1 \times D^3 \subset M_0 \times \{1\} \subset M_0 \times I$. This yields a 5-manifold $N_{1/2}$ with boundary $M_0 \cup M_{1/2} \cup P \times I$. Next attach 3-handles $D^3 \times D^2$ on embeddings $S^2 \times D^2 \subset M_{1/2}$, to get N with boundary $M_0 \cup M_1 \cup P \times I$. In the intermediate manifold $M_{1/2}$ there are two sets of embedded 2-spheres; the attaching spheres for the 3-handles, and the dual spheres of the 2-handles. The matrix of intersection numbers between these two sets represents the boundary homomorphism in the relative chain complex, so we want it to be the identity.

To be more specific about the 2-handles, suppose M_0 is a union of two 4-manifolds $V \cup W$ intersecting in a 3-manifold in their boundaries. Suppose there are framed embeddings of circles in the common boundary, denoted by $S^1_j \times D^2 \to V \cap W$. Collars on the boundaries in V and W extend this to an embedding $\bigcup_j S^1_j \times D^3 \to M_0$, which intersects V in $\bigcup_j S^1_j \times D^3_-$, where D^3_- denotes the lower half disk. Similarly the intersection with W is the product with the upper half disk D^3_+. Use these embeddings to attach 2-handles, written as $D^2_j \times (D^3_- \cup D^3_+)$. This gives a description of $M_{1/2}$ as $\left(V \cup (\bigcup_j D^2_j \times S^2_-)\right) \cup \left(W \cup (\bigcup_j D^2_j \times S^2_+)\right)$.

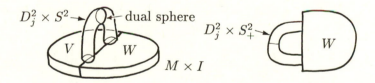

The dual spheres for the handles are $\{0\}_j \times S^2$, as shown. ∎

Proof of 6.1: Suppose $A \to M_0$ is an immersion of a union of transverse pairs, as described in the theorem. Define V to be a union of balls in M_0, one about the distinguished intersection point in the image of each transverse pair. W is the complement of the interior of V, and denote by $A_0 \to W$ the immersion of a union of disks defined by the part of A which lies in W.

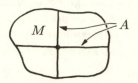 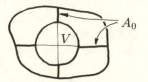

In a model for a transverse pair take parallel copies of the spheres. Push a piece of each parallel outside a neighborhood of the original, so it intersects the boundary of V in a circle. These circles divide the spheres into disks F and E, which can be chosen so that the F are disjointly embedded in the model, and the E intersect V only in the boundary, and each component of E intersects one of the original spheres in a single point. Note that the components of E are embedded, but there are intersections between them.

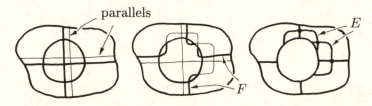

The immersion A carries these disks E to disks (also denoted E) in W, with boundaries denoted by $\{S_j^1\}$. Attach 2-handles on them as above, to get a handlebody with top $M_{1/2} = \left(V \cup (\bigcup_j D_j^2 \times S_-^2)\right) \cup \left(W \cup (\bigcup_j D_j^2 \times S_+^2)\right)$. A_0 defines an immersion of a union of disks into the second piece, and the disks E fit together with the core disks $D_j^2 \times \{+\} \subset D_j^2 \times (S_+^2)$ to give transverse spheres A_0^t.

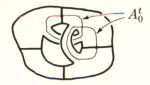

Now recall that according to the algebraic intersection hypothesis on A there are Whitney disks in M_0 for all intersections except the canonical transverse ones. Since $\pi_1 M_0 \simeq \pi_1\left(W \cup (\bigcup_j D_j^2 \times S_+^2)\right)$, there are Whitney disks for all intersections of A_0 in this latter manifold. Use the transverse spheres A_0^t to get disks disjoint from A_0.

Next construct transverse capped surfaces for the Whitney disks. Begin with the linking tori of intersection points, with normal disks as caps as in example 2 of 2.1, and use the transverse spheres A_0^t to get caps disjoint from A_0. We claim the resulting immersion of the capped surfaces is π_1-null in the complement of A_0. It lies in a neighborhood of A and the new 2-handles, which by hypothesis is π_1-null in M_0. It was observed above that $W \cup (\bigcup_j D_j^2 \times S_+^2)$ has the same fundamental group, and since A_0 has transverse spheres in this manifold, the complement of this again has the same fundamental group.

Since the Whitney disks have π_1-null transverse capped surfaces in the complement of A_0, theorem 5.2 applies to show there are disjointly embedded Whitney disks.

Use the embedded Whitney disks to construct a regular homotopy (rel boundary) of A_0 to an embedding in $W \cup (\bigcup_j D_j^2 \times S_+^2)$. Putting back $V \cup (\bigcup_j D_j^2 \times S_-^2)$ gives a regular homotopy of A in $M_{1/2}$ to an embedding A'. The final step is to add 3-handles in the complement of A' to finish construction of the s-cobordism.

The disks F in the model transverse pairs are now carried to embedded disks in $M_{1/2}$, which we will denote by $A'(F)$. These fit together with the disks $D_j^2 \times \{-\} \subset D_j^2 \times S_-^2$ to give framed embedded 2-spheres, disjoint from A'. Attach 3-handles on these to construct N. There is an immersion in N of the union of transverse pairs, crossed with I, constructed from $A \times I$ and the regular homotopy. The immersion in M_1 is A', an embedding, so the only thing left is to verify that N is an s-cobordism.

The spheres $A'(F) \cup D_j^2 \times \{-\}$ are regularly homotopic to the spheres $A(F) \cup D_j^2 \times \{-\}$, since A' is regularly homotopic to A. Since A is disjoint from the dual spheres of the 2-handles $\{0\}_j \times S^2$, and regular homotopy does not change intersection numbers, the algebraic intersections are the same as those of $D_j^2 \times \{-\}$ with $\{0\}_j \times S^2$, namely the identity matrix. This implies N is an s-cobordism, so completes the proof. ∎

6.3 Proof of 5.3

The objective here is to convert a disk embedding problem into one for transverse pairs. Suppose $A \colon (\bigcup_j D_j^2) \to M_0$ is an immersion with algebraically transverse spheres A^t, as in 5.3.

Let V be a collar of the boundary of M_0, and W the complement of the interior of this collar. Arrange for A to intersect V in a collar on its boundary, then it defines an immersion—denoted A_0—in W. Denote the boundary circles of A_0 by $S_j^1 \times D^2 \to V \cap W$. Attach 2-handles on these as in 6.2, to get a handlebody with top $\left(V \cup (\bigcup_j D_j^2 \times S_-^2)\right) \cup \left(W \cup\right.$

$(\bigcup_j D_j^2 \times S_+^2)$. The disks $D_j^2 \times \{-\}$ fit together with the collar on ∂A to give embedded disks in the first piece with the same framed boundary as A. The disks $D_j^2 \times \{+\}$ fit together with A_0 in the second piece to give framed immersed spheres.

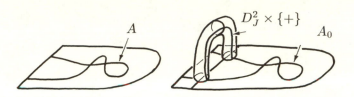

The spheres A^t and $A_0 \cup \bigcup_j D_j^2 \times \{+\}$ give an immersion of a union of transverse pairs in $(W \cup (\bigcup_j D_j^2 \times S_+^2)$ which satisfies theorem 6.1. Therefore there is an s-cobordism of this immersion to an embedding. Construct N by attachning this s-cobordism to the handlebody via $(W \cup (\bigcup_j D_j^2 \times S_+^2)$, and then attaching 3-handles on the embedded spheres s-cobordant to $A_0 \cup \bigcup_j D_j^2 \times \{+\}$.

First observe that N is an s-cobordism. For the purposes of simple homotopy the s-cobordism in the construction can be omitted, so N is equivalent to the original handlebody with 3-cells attached to $A_0 \cup \bigcup_j D_j^2 \times \{+\}$. The boundary homomorphism in the chain complex is the identity, so $M_0 \to N$ is a simple homotopy equivalence as required.

Next observe that this gives an s-cobordism from the immersion A to the embedding $\bigcup_j D_j^2 \times \{-\}$. For this we need an immersion of $(\bigcup_j D_j^2) \times I \to N$ with these immersions on the ends. There is an immersion of $(\bigcup_j D_j^2) \times I$ minus some balls, in the first handlebody, which is A on the bottom and $\bigcup_j D_j^2 \times \{-\} \cup (A_0 \cup \bigcup_j D_j^2 \times \{+\})$ on the top. The immersion in the s-cobordism, and the cores of the 3-handles fit together to fill in the missing balls.

The final ingredient needed is transverse spheres for the embeddings. For this we observe that the embeddings are π_1-negligible, and apply the criterion of 1.10. ∎

Exercise (1) Show that the two s-cobordism results are equivalent, by deducing 6.1 from 5.3, at least in the weakened form that half the spheres (one from each pair) are s-cobordant to π_1-negligible embeddings. (2) It is easy to see, using higher-dimensional methods, that an immersion in one end of an s-cobordism is s-cobordant to *some* immersion in the other end. Use this to deduce the full strength of 6.2 from the weakened version in the first exercise. ☐

6.4. Comments

The key step in the proof of 6.2 is the construction of the π_1-null transverse capped surface. The starting point—the linking torus and normal disks—is certainly π_1-null since it is constructed in a neighborhood of the intersection point. We then remove intersections with A, by adding copies of transverse spheres. Two spheres are required; one for the intersection point with each sheet. The result will be π_1-null if each sphere is, and either the spheres are disjoint, or loops through the intersection points are contractible.

In the proof given, some of the transverse spheres do intersect (see the picture in 6.3). The resulting loop in the transverse surface is homotopic to the loop through the original A intersection, so the transverse surface will be π_1-null if and only if the original immersion of A is. This is why we need the π_1-null hypothesis in 5.3 and 6.1.

To relax the π_1-null hypothesis on A seems therefore to require separating transverse spheres for A. The spheres occuring in 6.3 (for transverse pairs) have nontrivial algebraic intersection, so cannot be separated. Instead consider trying to use the technique to find Whitney disks for intersections between caps of a grope. In this case the transverse gropes for second stage surfaces can be used to get the normal disks disjoint from A. Disks from 2-handles in the s-cobordism can be used to get caps for these transverse gropes which are in a neighborhood of the body union handles $\bigcup_j D_j^2 \times (S_-^3)$, hence π_1-null. The problem is that the bottom stages of these transverse gropes intersect (see the exercise in 2.6).

The situation here is a little better than before, because the intersection is algebraically trivial. By changing one tranverse grope by 0-surgery (see the example in 2.1) we can obtain capped surfaces with disjoint bodies, but with caps of one intersecting the body of the other. This is not quite enough to separate them. In fact in this situation (second stages in a grope) it is not possible to get properly immersed transverse surfaces. This can be seen by interpreting the problem as finding surfaces in S^3 bounding links obtained by Bing doubling (see Chapter 12). It is classically known that Bing links are not boundary links.

These difficulties lead us to expect these embedding statements to be false without a π_1-null or poly-(finite or cyclic) hypothesis. Other approaches to the problem are described in Chapter 12.

The embedding up to s-cobordism results are sharpened forms of the surgery theorem of Freedman [6].

PART II

Applications to the Structure of Manifolds

The remainder of this book is devoted to consequences of the embedding theorem. Chapter topics are described briefly in the introduction, and in detail at the beginnings of the chapters. Here we introduce the term "good" for fundamental groups, and comment on the logic and significance of the material.

In Part II "good" is used to refer to fundamental groups for which the embedding theorem is known. In these terms the main theorem asserts that poly-(finite or cyclic) groups are "good." The intent is to allow easy identification of constraints imposed by the 4-dimensional topology, as distinct for example from special facts about the algebraic K-theory of poly-(finite or cyclic) groups. This terminology will also permit easy revision if the class of good groups is expanded.

The first three chapters (7–9) are concerned with basic technical results, whose analogs in other dimensions predate this development. We begin with several versions of the h-cobordism theorem in Chapter 7, and use this to establish existence and uniqueness results for smooth structures in Chapter 8. The smoothing theory (among other things) is applied in Chapter 9 to deduce topological transversality, handlebody structures, and existence and uniqueness of normal bundles.

There is a logical quirk in this development. The embedding theorem relies on a transversality result (the "immersion lemma" 1.2) to provide starting data. There is no independent proof of this for topological manifolds, so initially the theorem applies to give embeddings only in smooth manifolds. Embeddings in smooth manifolds are applied in Chapter 9 to prove the topological immersion lemma. After this the entire development of Part I also applies to topological manifolds. To avoid a logical circularity we indicate carefully which results are used in the proof of the immersion lemma, and verify that the manifolds can be given smooth structures.

Chapters 10 and 11 directly address classification questions for 4-manifolds. The reader with some background in high-dimensional manifold theory may want to begin with this material, and refer to earlier sections for technical details as needed.

The final Chapter, 12, describes some approaches to the question of whether the fundamental group hypothesis is necessary for the embedding theorem. At the time of writing this issue is unresolved, so the chapter serves as an introduction to current research on the subject.

CHAPTER 7

h-Cobordisms

The basic result is that certain 5-dimensional s-cobordisms are products. Section 7.1 concerns the classical case of compact h-cobordisms, while 7.2 and 7.3 deal with controlled and proper h-cobordisms respectively. There are refinements which give more information on the singularities of the product structures. These are used in the next chapter, in studying smooth and handlebody structures, while later chapters will primarily use the basic product result.

7.1. Compact s-cobordisms

Recall that an *h-cobordism* (rel boundary) is a manifold W with $\partial W = M_0 \cup M_1$ a union of submanifolds, such that W deformation retracts to each M_i. A compact h-cobordism has a Whitehead torsion $\tau(W, M_0) \in \mathrm{Wh}(\pi_1 W)$ defined (see Cohen [1], Milnor [2]), and an *s-cobordism* is defined to be an h-cobordism with vanishing torsion. In the controlled and proper cases more elaborate "torsions" are defined, and the vanishing of these is used to define s-cobordisms.

The product s-cobordism beginning with a manifold M is $M \times I$ with $\partial_0(M \times I) = M \times \{0\}$ and $\partial_1(M \times I) = (\partial M \times I) \cup (M \times \{1\})$. The classical s-cobordism theorem asserts this is the only one: if the dimension of an s-cobordism W is at least 6 then it is isomorphic to the product $M_0 \times I$. This theorem is "category independent" in the sense that if W is smooth, or PL, then the isomorphism with is also smooth or PL. The 5-dimensional version is weaker in two regards; there is a fundamental group restriction, and it only works topologically.

7.1A Theorem. *A compact s-cobordism of dimension 5 with good fundamental group has a topological product structure.*

Whitehead groups of poly-(finite or cyclic) groups are known to be primarily determined by finite subgroups (Quinn [9]). In particular if the group is torsion free the Whitehead group vanishes (Farrell and Hsiang [1]), and every h-cobordism is a product.

It is a standard observation that the theorem implies a characterization of h-cobordisms: fix a 4-manifold M, then homeomorphism classes (rel M) of h-cobordisms (W, M) correspond bijectively to the Whitehead group of $\pi_1 M$. The standard construction of h-cobordisms (Rourke and Sanderson [1, p.90]) works in this dimension, and shows there is an h-cobordism with any given torsion. The invariant is therefore onto $\mathrm{Wh}(\pi_1 M)$. Conversely if the fundamental group is good then this together with the theorem implies that h-cobordisms with the same torsion are homeomorphic.

As an application we give a proof of the 4-dimensional version of the "Poincaré conjecture."

7.1B Corollary. (Freedman [2]) *A 4-manifold homotopy equivalent to S^4 is homeomorphic to S^4.*

Proof: Let M denote the homotopy sphere, then the cone cM is a 5-manifold except possibly at the cone point. At the cone point it is a homology manifold, and this point is 1-LC embedded (see 9.3B). The flattening theorem for points (the high-dimensional analog of 9.3B, due originally to Bryant and Lacher [1]) implies that the cone is in fact a manifold at this point also. Delete an open ball from the cone to get a topological h-cobordism from M to S^4, and apply the theorem. ∎

We caution that the h-cobordism is not known a prori to have a smooth structure, so this requires the topological version of the theorem. This version depends logically on the developments of the next two chapters. The original proof (Freedman [2]) is longer, but avoids this dependence: delete a point from the 4-manifold and find a smooth proper h-cobordism of the result to $\mathbf{R}^4$. Then apply the smooth version of the proper h-cobordism theorem (7.3), which is immediately available.

The following slight extension of 7.1A is used in circumstances where fundamental group images can be arranged to be good, but the groups are not good, or the manifold is not compact.

An h-cobordism $(W, \partial_0 W)$ is said to "have a product structure on the complement" of $V \subset W$ if there is an isomorphism $\mathrm{cl}(W - V) \simeq \mathrm{cl}(\partial_0 W - \partial_0 V) \times I$, and V intersects this in $(\mathrm{cl}(\partial_0 W - \partial_0 V) \cap V) \times I$. We think of W as an "enlargement" of V, in a sense similar to the definition in the exercise after 5.1A.

7.1C Proposition. *Suppose W is a 5-dimensional h-cobordism which has a product structure in the complement of a compact s-cobordism V, and $\pi_1 V \to \pi_1 W$ has good image. Then there is a topological product structure on W which agrees with the given one off some compact set.*

The final statement in the section is the technical version which will actually be proved. The hypotheses are more complicated because initially we lack several basic facts about topological manifolds (eg. handlebody structures). The conclusions are more complicated because we detail how close we can come to getting a smooth product structure. Let $\mathcal{M}$ denote one of the categories of manifolds; DIFF, PL, or TOP.

7.1D Theorem (Technical version). *Suppose the immersion lemma is valid in $\mathcal{M}$, and (W, M) is a compact 5-dimensional s-cobordism with good fundamental group and an $\mathcal{M}$-handlebody structure rel M. Then there is a homeomorphism $M \times I \to W$, which is an $\mathcal{M}$ isomorphism except on a set $U \times I$, where U may be taken to be either a topological regular neighborhood of a 1-complex, or an $\mathcal{M}$-regular neighborhood of an $\mathcal{M}$ 2-complex K with the following negligibility property: any $\mathcal{M}$ immersed surface with boundary disjoint from K is $\mathcal{M}$ regularly homotopic rel boundary to an immersion disjoint from K.*

The proof is long, so before beginning it we explain the relation to 7.1A and 7.1C. In the categories $\mathcal{M}=$ DIFF, PL, the immersion lemma and existence of handlebody structures are known independently of the present development. This implies there are topological product structures on smooth or PL s-cobordisms. This fact (or rather the controlled version 7.2) is used in Chapter 9 to prove the immersion and handlebody lemmas in the topological category. These then imply that 7.1D applies to $\mathcal{M} = $ TOP, proving 7.1A.

Proposition 1.7C requires a slight extension of the proof. To get begin with the s-cobordism V and follow the proof to the point where the embedding theorem is used. The theorem is applied to find Whitney disks for intersections between boundary spheres of handles, in a level set M in a handlebody structure on V. The handlebody structure on $(V, \partial_0 V)$ extends to one on $(W, \partial_0 W)$ with the same handles. This therefore gives an enlargement of M, in the sense of the exercise after 5.1A, but introduces no new spheres and no new intersections. The fundamental groups of the level sets are the same as those of V and W, so by hypothesis the image of $\pi_1 M$ in the enlargement is good. Exercise 5.1A therefore applies to give embedded disks in the larger manifold. The proof now continues as before, in the larger h-cobordism W.

Proof of 7.1D: Begin with an $\mathcal{M}$ handlebody structure on (W^5, M). The standard proof of the high dimensional result (Rourke-Sanderson [**1**, §6], Milnor [**3**, §8]) shows that the handlebody structure can be deformed to one with no 0- or 1-handles, and dually no 4- and 5-handles. This leaves handles only in dimensions 2 and 3. The cellular chain complex based on this, with $\mathbf{Z}[\pi_1 W]$ coefficients, therefore has chains only in dimension 2 and 3. Further, the 2- and 3-handles can be arranged and enumerated so that the boundary homomorphism in the chain complex is the identity matrix.

We interpret this geometrically. W is $\mathcal{M}$-isomorphic to $M \times I \cup (H_1^2 \cup$

$\cdots \cup H_n^2) \cup (H_1^3 \cup \cdots \cup H_n^3)$. Let N denote the level between the 2- and 3-handles, then there are two collections of framed embedded 2-spheres; the attaching maps of the 3-handles $A_1, \ldots A_n$, and the dual spheres of the 2-handles $B_1, \ldots B_n$. The chain complex condition is that the algebraic intersection number $\lambda(A_i, B_j) = 0$ if $i \neq j$, and $= 1$ if $i = j$.

The dimension restriction in the standard proof occurs in the next step; general position is used to find embedded Whitney disks for the algebraically cancelling intersections between A and B. The isotopy of A obtained by pushing across these disks gives a further deformation of the handle structure to one whose handles can be cancelled. Cancellation leaves only the product $M \times I$, as required for the theorem.

In dimension 5 the following modifications are required: first an isotopy of A which ensures that the total collection of spheres $A \cup B$ has transverse spheres. These are used to find immersed Whitney disks for the extra intersections, with interiors disjoint from $A \cup B$. The main embedding theorem is used to replace these with topologically embedded disks. Pushing across these disks gives a handle structure whose 3-handles are attached topologically, and cancelling handles yields a topological product structure on W. The proof is completed by analysing the damage the topological move does to the product structure.

Suppose, then, that W has dimension 5. The first step is to observe that the collections A, B separately have transverse spheres in N. N can be described as obtained from M by deleting the interiors of the attaching maps of the 2-handles $H_i(S^1 \times D^3)$, and adding $H_i(D^2 \times S^2)$. B_i is the sphere $H_i(\{\text{point}\} \times S^2)$, which intersects the disk $H_i(D^2 \times \{\text{point}\})$ in a single point. Transverse spheres can therefore be constructed from these disks and immersed disks in $M - \text{int}H_i(S^1 \times D^3)$ intersecting the boundary in $H_i(S^1 \times \{\text{point}\})$.

$H_i(S^1 \times \{\text{point}\})$ is nullhomotopic in $M \times I \cup (H_1 \cup \cdots \cup H_n)$, and the composition $M \to M \times I \cup (H_1 \cup \cdots \cup H_n) \to W$ is an isomorphism on π_1, so $H_i(S^1 \times \{\text{point}\})$ is nullhomotopic in M. By general position it is also nullhomotopic in the complement of the attaching circles of the 2-handles. Applying the immersion lemma (in $\mathcal{M}$) to such nullhomotopies yields the desired immersed disks in $M - \text{int}H_i(S^1 \times D^3)$.

The spheres produced in this way intersect the B spheres in single points, but may not be framed (see 1.3). A little more care in choosing the nullhomotopy gives spheres which can be framed after twisting (again see 1.3), but unframed spheres will be adequate here. (*Exercise*: show in fact that B has framed *embedded* transverse spheres in N.)

Applying the same argument to the dual handlebody produces unframed transverse spheres for the spheres A.

The next step is to show that A can be changed by $\mathcal{M}$ isotopy—in fact finger moves through B—so that the new $A \cup B$ has transverse spheres (framed ones this time). Begin by applying isotopies from part (2) of the immersion lemma to make the $A \cap B$ intersections standard. Let A', B' denote parallel copies of A, B, then according to the algebraic intersection hypotheses these are algebraically transverse spheres for $A \cup B$. Choose (again using the immersion lemma) immersed Whitney disks V_* for the extra $A' \cap B$ intersections, and W_* for the extra $B' \cap A$ intersections.

Make the interior of V disjoint from B by sums with the unframed transverse spheres for B, and correct the framing (if necessary) by spinning about the edge on A'. Similarly make the interior of W disjoint from A by sums with the unframed transverse spheres, and correct the framing by spinning about the edge on B'. Next push A off V by finger moves through B, and B off W by finger moves through A. These finger moves can all be done disjointly (and in the category $\mathcal{M}$). Also the inverse of a finger move of B through A is a finger move of A, so we may consider these as an isotopy of A alone. Finally since these take place in neighborhoods of arcs, they can be done disjointly from A' and B'.

The result is a collection of immersed Whitney disks with interiors disjoint from $A \cup B$. Pushing $A' \cup B'$ across these eliminates the extra intersections, and gives transverse spheres for $A \cup B$.

We now construct Whitney disks for the extra intersections of $A \cap B$. Begin with immersed ones of category $\mathcal{M}$ (using the immersion lemma), and use sums with the transverse spheres to make the interiors disjoint from $A \cup B$. There are algebraically transverse spheres for these disjoint from $A \cup B$; use the transverse spheres to move the caps of linking tori off $A \cup B$, and contract (see example 2 of 2.1). Finally the transverse spheres imply the complement of $A \cup B$ has the same fundamental group as M, which is assumed to be good. Therefore the main embedding theorem 5.1 implies that there are topologically embedded Whitney disks.

More precisely, the first part of the proof of 5.1 replaces the immersed Whitney disks by properly immersed capped towers T_* of category $\mathcal{M}$, and with transverse spheres disjoint from $A \cup B$. The second part produces, inside an $\mathcal{M}$ regular neighborhood of T, topologically embedded disks W_*.

The product structure on the s-cobordism is obtained by pushing A across the W_* and cancelling the handles, as indicated above. This structure can fail to be of category $\mathcal{M}$ only on the attaching region of the handles in M, union with the neighborhood of the W_* in which the pushing isotopy takes place. This is contained in the attaching region union with an $\mathcal{M}$-regular neighborhood of the 2-complex T. The next step is to analyse these sets.

By the "attaching region in M" we mean the intersection of $M \times \{1\}$ and $\cup_i H_i^2 \cup \cup_i H_i^3$ in the handlebody. This can be pictured as obtained from a neighborhood of $A \cup B$ in N by "undoing" the 2-handles; removing the B spheres $S^2 \times D^2$ and replacing them by neighborhoods of circles $D^3 \times S^1$. The A spheres get holes cut in them at intersections with B, and the edges of these holes are parallel to the new circles. The attaching region is therefore a regular neighborhood of a 2-complex obtained by puncturing spheres and identifying the edges to circles. We will denote this 2-complex by K_0.

For the topological analysis we are interested in a neighborhood of $A \cup B \cup W_*$. Let A' denote the spheres obtained by pushing A across W_*. A can be recovered from A' (up to isotopy rel B) by finger moves along certain arcs c_*. A regular neighborhood of $A \cup B \cup W_*$ in N can therefore also be described as a regular neighborhood of $A' \cup B \cup c_*$.

Now pass to M by replacing B by circles. The A' have the property that A_i' intersects only B_i, and in only one point. The 2-complex obtained by cutting holes at the intersections and identifying edges corresponding to each B_i is therefore a disjoint union of disks. A regular neighborhood of a union of disks and arcs is a regular neighborhood of a 1-complex. This verifies the assertion that the non-$\mathcal{M}$ behavior is confined to a topological regular neighborhood of a 1-complex in M.

We now get an $\mathcal{M}$ description of the problem region. Note that since the original handles are $\mathcal{M}$, the attaching region in M is an $\mathcal{M}$-regular neighborhood of the 2-complex K_0 defined above. The Whitney disks lie in an $\mathcal{M}$-regular neighborhood of the $\mathcal{M}$-capped tower T, so the problem region is contained in an $\mathcal{M}$-regular neighborhood of the 2-complex $K_0 \cup T \subset M$. We show that this 2-complex has the regular homotopy property described in the theorem.

Suppose $S \to M$ is an $\mathcal{M}$ immersion of a surface, with boundary disjoint from $K_0 \cup T$. By general position we can change S by isotopy to be disjoint from the circles in K_0, and then can think of it as a map $\hat{S} \to N$ disjoint from B. If we change $S \to M$ by an isotopy which passes through the circle corresponding to B_i, then $\hat{S}$ changes by connected sum with a parallel of B_i.

The next step is to arrange that the intersections $\hat{S} \cap A$ occur in pairs with Whitney disks (immersed, in N). Suppose some intersections have Whitney disks chosen, but there is an intersection $\hat{S} \cap A_i$ which does not have one. Push S through the circle corresponding to B_i, then $\hat{S}$ changes by connected sum with a parallel of B_i. All but one of the $B_i \cap A$ intersections are in pairs with Whitney disks (in fact Whitney towers T_*), so parallels of these provide Whitney disks for all but one of the new $\hat{S} \cap A$ intersections. The remaining new intersection is in $\hat{S} \cap A_i$. If the push of S through the circle takes place along an appropriate arc, and with proper orientation, then this new intersection can be paired with the old $\hat{S} \cap A_i$ intersection, with a Whitney disk. This construction reduces the number of unpaired intersections, so repetition leads to the desired conclusion.

$\hat{S}$ can be made disjoint from the towers T by pushing down and off across the edge on A. This introduces new $\hat{S} \cap A$ intersections, but since they arise from finger moves they are arranged in pairs with Whitney disks.

Use sums with the transverse spheres for $A \cup B$ to get Whitney disks with interiors disjoint from $A \cup B$. Similarly, since the towers T_* have transverse spheres disjoint from $A \cup B$ we can get Whitney disks disjoint from T.

Now define a regular homotopy of $\hat{S}$ by immersed Whitney moves across these Whitney disks. This takes place in the complement of B, so defines a regular homotopy of S in M. The new $\hat{S}$ is disjoint from $A \cup B \cup T$ in N, so the new S is disjoint from the 2-complex $K = K_0 \cup T$ in M.

This completes the verification that $K_0 \cup T$ has the regular homotopy property, and completes the proof of 7.1D. ∎

Exercise   Suppose W is a compact 5-dimensional s-cobordism of category $\mathcal{M}$ and arbitrary fundamental group. Define a *sum stabilization* to be the s-cobordism obtained by deleting a regular neighborhood of an arc from M_0 to M_1, and subsituting $(S^2 \times S^2 - \text{ball}) \times I$. Denote this by $W \#_I (S^2 \times S^2 \times I)$, and note it changes M_0 and M_1 by connected sum with $S^2 \times S^2$. If W has a handlebody structure, and the arc is chosen to be a collar arc which misses the handles, then this operation also changes intermediate levels by connected sum. Prove: there is an $\mathcal{M}$ product structure on $W \#_I k(S^2 \times S^2 \times I)$, for some k. (See the exercise in 1.9, and the comments following.) □

7.2. Controlled h-cobordisms

The objective here is to get information about the size of product structures on an s-cobordism, as measured in a control space. An introduction to the high-dimensional theory is given in Quinn [7], see also Quinn [3] and Chapman [1].

Suppose X is a metric space, and $\delta: X \to (0, \infty)$ is continuous. An δ-h-*cobordism* is, as before, a manifold W with $\partial W = M_0 \cup M_1$, such that W deformation retracts to each M_i. Further there is a map $W \to X$, and the images of the deformation retractions have radius less than δ in X. Generally, a homotopy $F: W \times I \to X$ has radius $< \delta$ if the arc $F(\{w\} \times I)$ lies in the ball of radius $\delta(F(w, 0))$ about $F(w, 0)$. Finally, we require $W \to X$ to be proper (preimages of compact sets are compact; W itself is not required to be compact).

A product structure $M_0 \times I \simeq W$ defines—by composition—a homotopy in X. We say the product structure has radius less than δ if both this homotopy and the one obtained by reversing the I coordinate have radius less than δ.

As in section 5.5 we use a reference map $p: E \to X$ to specify local fundamental groups. Here we will also need to insure these local groups do not change too wildly, by requiring p to be a simplicial neighborhood deformation retract (NDR). This means there is a commutative diagram

$$
\begin{array}{ccc}
E & \xrightarrow{\ c\ } & K \\
\downarrow{\scriptstyle p} & & \downarrow \\
X & \xrightarrow{\ c\ } & L
\end{array}
$$

with $K \to L$ simplicial, and deformation retractions of neighborhoods to the images of E, X so that the deformation in K covers the one in L. The definition of $W \to E$ being δ-1-connected (over X) is given in 5.4.

7.2A Theorem (Controlled h-cobordism). *Suppose $p: E \to X$ is a simplicial NDR with good local fundamental groups. Given $\epsilon: X \to (0, \infty)$ there is $\delta: X \to (0, \infty)$ so that if $(W^5; M_0, M_1) \to E$ is a δ-1-connected δ-h-cobordism over X then there is a topological product structure of diameter less than ϵ on W if and only if the controlled torsion $q_1(W, M_0) \in H_1^{lf}(X; \mathcal{S}(p))$ vanishes.*

We discuss the obstruction, and begin by assuring the reader that it is not necessary to understand it in any detail in order to use the theorem effectively. In fact in nearly every application we make of it the obstruction group is trivial (see the corollary below).

The obstruction group is a locally finite homology group with coefficients in a "cosheaf of spectra." This and the invariant q_0 are defined in Quinn [**3**], where the high-dimensional analog of the theorem is proved. These homology groups have many useful formal properties, and in favorable cases can be computed using a spectral sequence. This spectral sequence has as input homology with coefficients in Whitehead groups and lower algebraic K-theory groups of local fundamental groups, see Quinn [**3**, §1, 8] for details and examples. In particular if the K and Whitehead groups vanish for all fundamental groups $\pi_1(p^{-1}(x))$, where $x \in X$, then the spectral sequence disappears, and the obstruction group is trivial.

The class of poly-(finite or cyclic) groups (for which the 4-dimensional topology is known to work) is somewhat favorable here since quite a bit is known about the K-theory (Farrell-Hsiang [**1**], Quinn [**9**]). In particular the appropriate K and Whitehead groups vanish if the fundamental groups are torsion free. This implies vanishing of the controlled h-cobordism obstruction groups, and yields the following corollary:

7.2B Corollary. *Suppose* $p \colon E \to X$ *is as above, with torsion free poly-(finite or cyclic) local fundamental groups, and* ϵ *is given. Then there is* δ *such that every 5-dimensional* δ-1-*connected* δ-*h-cobordism* $(W^5; M_0, M_1) \to E$ *has an* ϵ *product structure.*

As in the compact case we prove a technical refinement, which is primarily needed in the next chapter. This includes information about non-differentiability and is also relative with respect to subsets of X. By "relative" we mean that open sets $V \subset U \subset X$ are given, we consider $(W, M_0) \to X$ with product structure already given "over V", and want to extend this to a product structure "over U".

A product structure "over U" on $r \colon (W, M_0) \to X$ is an open set $M_0 \supset N \supset (r^{-1}U) \cap M_0$ and an open embedding $N \times I \to W$ which is the identity on $N \times \{0\}$ and contains $r^{-1}U$. In trying to construct these, it is appropriate only to impose hypotheses over U. Accordingly, we say r is an δ-h-cobordism "over U" if $r^{-1}(U)$ has deformation retractions rel M_i in W into M_i, of radius less than δ in X.

There is a little slippage which we allow for as follows: define V^δ to be the set of points $\{y : d(y, v) < \delta(y)\}$. Then we will be given a product structure over V^δ, will restrict it to V, and then extend this restriction to a product structure over U. More precisely, the statement is in terms of handlebody structures, the connection with products being that a handle structure with no handles is a product.

As in 7.1D, let $\mathcal{M}$ denote one of the categories of manifolds; DIFF, PL, or TOP.

7.2C Theorem (technical controlled h-cobordism). *Suppose the immersion lemma is valid in $\mathcal{M}$, and $p\colon E \to X$ is a simplicial NDR, with good local fundamental groups. Suppose $V \subset U \subset X$ are open and $\epsilon\colon X \to (0, \infty)$ is given. Then there is $\delta\colon X \to (0, \infty)$ so that if $(W^5; M_0, M_1) \to E$ is a δ-1-connected δ-h-cobordism over $U^{2\epsilon}$ with a $\mathcal{M}$-handlebody structure rel M of diameter less than δ and no handles over $V^{2\epsilon}$, then there is an obstruction $q_1(W, M_0) \in H_1^{lf}(U^\epsilon - V^\epsilon; \mathcal{S}(p))$. If this vanishes there is a topological handlebody structure of diameter less than ϵ with no handles over U, which agrees with the given structure except over $U^{2\epsilon} - V$. Further, the resulting product structure over U is an $\mathcal{M}$ isomorphism except on a set $W \times I$, where W may be taken to be either an ϵ topological regular neighborhood of a 1-complex, or an ϵ-$\mathcal{M}$-regular neighborhood of an $\mathcal{M}$ 2-complex K with the following negligibility property: any $\mathcal{M}$ immersed surface with boundary disjoint from K is ϵ, $\mathcal{M}$ regularly homotopic rel boundary to an immersion disjoint from K.*

By "ϵ regular neighborhood" we mean one with a mapping cylinder structure such that the images of the mapping cylinder arcs in X have diameters less than ϵ. The point is that if something is disjoint from K, then it can be made disjoint from a regular neighborhood of K by isotopy pushing along mapping cylinder arcs. If these arcs have diameter less than ϵ then this isotopy also has diameter less than ϵ.

Proof: As in 7.1 the proof is sketched, indicating the modifications required in dimension 4. We also subsitute the word "control" for phrases like "given $\delta_n > 0$ there is $\delta_{n+1} > 0 \ldots$," and "small" for "diameter less than δ_n, appropriate n." Little is lost, since it is usually a routine matter to reconstruct the more complicated statement, and the brief versions are much clearer. (See for example the proof of Theorem 5.4).

Begin with a small $\mathcal{M}$ handlebody structure on (W, M_0). Since it is a controlled h-cobordism over $U^{2\epsilon}$ this structure can be manipulated with control to have only 2- and 3-handles over U^ϵ (Quinn [**2**, 6.3]). The obstruction homology class is identified in Quinn [**3**, 6.2 and §3] as an equivalence class of the intersection matrix between these handles. Therefore the vanishing of the obstruction implies that the handlebody can be further manipulated (with control) so that the algebraic intersections are the identity matrix.

As in the proof of 7.1C the intersection matrix is interpreted as intersections between attaching spheres A_i and dual spheres B_i in an intermediate level. The spheres A can be moved as in 7.1D (with control) so that the collection $A \cup B$ has small transverse spheres. These are used as in 7.1D to construct small immersed Whitney disks for the extra intersections $A \cap B$. Since we have assumed that the local fundamental groups are good, the controlled embedding theorem 5.4 gives small disjoint topologically embedded Whitney disks for these extra intersections.

Pushing the attaching maps A across these Whitney disks gives a small topological handlebody structure such that all the handles over U can be cancelled. Cancelling them leaves a handlebody structure satisfying the conclusions of the theorem.

The singularities introduced into the $\mathcal{M}$ structure by the topological move are analysed in exactly the same fashion as in the proof of 7.1D. ∎

7.3. Proper h-cobordisms

A *proper* h-cobordism is a triple $(W; M_0, M_1)$ which deformation retracts to both M_0 and M_1, as above. W is not required to be compact, but the deformations are required to be proper. We will recognize some proper h-cobordisms as controlled h-cobordisms over a graph, so 7.2A and 7.2B give criteria for product structures.

If W is locally compact we can define a "component of the end" of W to be an equivalence class of a decreasing sequence of open sets U_i. Each U_i is a nonempty component of the complement of some compact set, and if $K \subset W$ is compact then for some i, $K \cap U_i = \phi$. Two sequences U, U' are equivalent if for each i there is j such that $U_i \supset U'_j$ and $U'_i \supset U_j$. We denote the set of components of ends of W by eW. In the endpoint compactification notation of section 3.8 this is $E(W) - W$.

Let opencone(eW) denote the open cone $[0, \infty) \times (eW)/\{0\} \times (eW)$, with vertex denoted by 0. If W has only finitely many ends then there are proper maps $W \to$ opencone(eW) obtained by choosing a sequence U_i^α for each component α, and mapping U_i^α to $\{\alpha\} \times (i, \infty)$. Further, if W is a proper h-cobordism and δ: opencone$(eW) \to (0, \infty)$ is continuous, then there is such a map which makes W a δ-h-cobordism. Roughly this is obtained by using compactness and properness to choose sequences so that the deformations keep U_{i+1}^α in U_i^α, and $W - U_i^\alpha$ in $W - U_{i+1}^\alpha$. Then map U_i^α to $\{\alpha\} \times (n_i, \infty)$, where the n_i are within δ of each other. Details for this argument appear in the proof of 9.3B: the $[0, \infty)$ coordinate part of the demonstration that the W there can be made a δ-h-cobordism over $N \times [0, \infty)$.

The next objective is a condition which allows control of the local fundamental groups in such a δ-h-cobordism. We say that fundamental groups are *stable* at an end α if there is a defining sequence such that the image of $\pi_1 U_{i+1} \to \pi_1 U_i$ does not depend on i. (By this we mean that if $i < j$ then (image)$_j \to$ (image)$_i$ is an isomorphism.) Denote this image by $\pi_1 \alpha$.

The inclusions $U_i \subset W$ define a homomorphism $\pi_1\alpha \to \pi_1 W$. Let E denote the open mapping cylinder of the corresponding maps of Eilenberg-MacLane spaces $K(\pi_1\alpha, 1) \to K(\pi_1 W, 1)$, and let p denote the natural projection $E \to \text{opencone}(eW)$. The map of W to opencone(eW) factors through p, and $W \to E$ can be arranged to be δ-1-connected for any given δ (still presuming the ends of W have stable fundamental groups).

Summing up, we have:

7.3A Lemma. *Suppose $(W; M_0, M_1)$ is a proper h-cobordism with finitely many ends, with stable fundamental groups, and suppose a function δ: opencone$(eW) \to (0, \infty)$ is given. Then there is a $(\delta, 1)$-connected map $W \to E$ which makes W a (δ, h)-cobordism over opencone(eW).*

Combining this lemma with 7.2 gives proper h-cobordism theorems for h-cobordisms with finitely many stable ends. A more elaborate approach, developed by L. Siebenmann [1] in higher dimensions and implemented in dimension 4 in Freedman [2], applies when there are infinitely many ends, possibly not stable.

Combining the lemma with 7.2B yields:

7.3B Corollary. *Suppose $(W; M_0, M_1)$ is a 5-dimensional proper h-cobordism with finitely many ends, with stable fundamental groups, and the fundamental groups of W and each end are torsion-free poly-(finite or cyclic). Then W has a topological product structure.*

Combining the lemma with the relative version of the controlled theorem, in a neighborhood of a single end, gives:

7.3C Corollary. *Suppose $(W; M_0, M_1)$ is a 5-dimensional proper h-cobordism in a neighborhood of a single end α, which has stable fundamental group which is good. Then there is an obstruction in $\tilde{K}_0(\mathbf{Z}[\pi_1\alpha])$ whose vanishing is equivalent to the existence of a product structure on W in a neighborhood of α*

Proof: According to the lemma we can find a map of a neighborhood of the end to $[0, \infty)$ which makes it a δ-h-cobordism over some neighborhood of the end. Let $U = (1, \infty)$, then 7.2C gives an obstruction in $H_1^{lf}((1, \infty) \times eW; \mathcal{S}(p))$. This locally finite homology group is isomorphic to $H_0(\text{pt}; \mathcal{S}(p)) = \tilde{K}_0(\mathbf{Z}[\pi_1\alpha])$, as required. ∎

7.4. References

The simply-connected version of 7.1 appeared in Freedman [2], and was extended to poly-(finite or cyclic) fundamental groups in Freedman [4]. More precisely, these give a proof of 7.1D, with the immersion lemma and handlebody hypotheses avoided by assuming W has a smooth structure. The proof of 7.1A was completed by the proofs of the handlebody and immersion hypotheses in Quinn [4]. The sum-stable product theorem in the exercise in 7.1 comes from Quinn [1].

In higher dimensions the h-cobordism theorem applies without regard to category; a smooth, etc. s-cobordism has a smooth, etc. product structure. The nonexistence of smooth product structures on some simply-connected 5-dimensional s-cobordisms follows from Donaldson [2]; see the next chapter. Cappell and Shaneson [2], [3], have shown that there are 4-dimensional s-cobordisms (so the ends are 3-dimensional) which do not have topological product structures.

The simply-connected version of 7.2 appeared in Quinn [4]. Quinn [7] provides some introductory material on controlled h-cobordisms. A somewhat different approach to controlled h-cobordisms, which can also be modified to apply in dimension 5, is given in Chapman [1]. A special case in dimension 5, sufficient to prove the 4-dimensional annulus conjecture, is discussed in Edwards [1].

CHAPTER 8

Smooth structures

The main results of this chapter are partial existence and uniqueness results for smooth structures on 4-manifolds. These are limited by
smooth nonexistence results obtained from differential geometry, so in
fact their main use (in the next Chapter) is to obtain further information about the topological category. We have included discussions of the
4-dimensional smooth nonexistence results, and—to make the idiosyn-
crasies of this dimension clear—a description of the smoothing theory
in other dimensions. In 8.6 and 8.7 appropriate sum-stable and bundle
versions are shown to follow the high dimensional pattern. Finally in
8.8 smooth structures in the complement of points are considered.

8.1 Uniqueness of smoothings

The first question considered is uniqueness; if $M \to N$ is a homeomorphism of smooth 4-manifolds, is it isotopic to a diffeomorphism? The
answer is usually "no", so a more limited question is necessary for positive results. We suppose $K \subset M$ is a smooth subcomplex of M, and ask
if there is an isotopy to a homeomorphism which is a diffeomorphism on
a neighborhood of K. The answer is still usually "no", but the following
is true:

8.1A Theorem. (The "Hauptvermutung," or "annulus conjecture")
*Suppose $h: M \to N$ is a homeomorphism of smooth 4-manifolds which is
a diffeomorphism on the boundary, and $K \subset M$ is a smooth subcomplex.*

(1) *If $\dim K \leq 1$ then h is isotopic rel boundary to a homeomorphism
which is a diffeomorphism on a neighborhood of K, and*

(2) *if $\dim K = 2$ then K can be changed rel boundary by either
topological isotopy or smooth regular homotopy to K' so that h
is isotopic to a homeomorphism which is a diffeomorphism on a
neighborhood of K'.*

*Further, if $\epsilon: N \to (0, \infty)$ and a neighborhood U of K are given, then
there is such an isotopy which is constant off U and has diameter less
than ϵ.*

This theorem, and the proof given, are from Quinn [**4**], [**6**].

We will see in 8.4D that the hypotheses given are necessary; it is generally impossible to smooth h near a 2-dimensional K without changing it as in (2), and impossible to smooth near a 3-dimensional K even allowing such changes. A notation will be convenient: we define a *displacement* of K to be the image of K under a homeomorphism of M topologically isotopic to the identity. In these terms statement (2) asserts that a homeomorphism can be smoothed in a neighborhood of a displacement of a 2-complex.

In (2), recall that a regular homotopy is a sequence of Whitney moves, finger moves, or ambient isotopies. In fact the regular homotopy produced in the proof requires only finger moves and isotopies.

Proof: Two results from outside our development will be used: the 5-dimensional case of a high-dimensional smoothing theorem, which will be described in 8.3B, and the following "approximation implies isotopy" theorem:

8.1B Lemma. (A. V. Cernavskii; Kirby and Edwards [1]) *Suppose $h\colon M \to N$ is a homeomorphism of manifolds (of any dimension), and $K \subset M$ is closed, U is a neighborhood of K, and $\epsilon\colon N \to (0, \infty)$ is given. Then there is $\delta > 0$ so that if $g\colon U \to N$ is a homeomorphism on its image, and is within δ of h on U, then there is an isotopy constant off U and of diameter less than ϵ, of h to h' so that $h' = g$ on K.*

This reduces the proof of the theorem to the construction of a good approximation g; the isotopy to h is automatic.

Suppose K is a smooth subcomplex, and let U_0 be a regular neighborhood. Let U_1 be the image $h(U_0) \subset N$, and define W to be the mapping cylinder of $h\colon U_0 \to U_1$. W is a topological 5-manifold with a smooth structure on its boundary $U_0 \cup U_1$. The obstruction to extending the smooth structure to one on W lies in $H_4(W, U_0 \cup U_1; \mathbf{Z}/2)$ (see 8.3B). Since h is a homeomorphism, W is topologically a product $U_0 \times I$. Consequently this group is isomorphic to $H_3(U_0; \mathbf{Z}/2) = H_3(K; \mathbf{Z}/2)$, which vanishes since K is at most 2-dimensional. Therefore there is such a smooth structure on W.

Consider the projection $W \to U_1$. Topologically this is a projection of a product $U_1 \times I$ to the I coordinate, so in particular is a δ-1-connected δ-h-cobordism over U_1 for every $\delta\colon U_1 \to (0, \infty)$. Since it has a smooth structure the controlled h-cobordism theorem 7.2C applies to give a new product structure, and in particular a homeomorphism $U_0 \to U_1$, within any given ϵ of the original. Further this homeomorphism is a diffeomorphism except on a set $W \subset U_0$.

According to 7.2C, the singular set W can be chosen to lie in a small smooth regular neighborhood of a smooth 2-complex with a certain regular homotopy property. If the complex K of 8.1 has dimension 1, then by general position there is a small smooth isotopy which moves it disjoint from this 2-complex and its regular neighborhood, and therefore the singular set. Or, composing the homeomorphism $U_0 \to U_1$ with the inverse of this isotopy yields a homeomorphism smooth near K. Since this can be done arbitrarily close to the original homeomorphism, the lemma provides the isotopy required to complete case (1) of the theorem.

When the dimension of K is 2, the regular homotopy property of the singular set described in 7.2C yields the smooth regular homotopy conclusion of statement (2). Alternatively, 7.2C asserts that the singular set is contained in a small topological regular neighborhood of a topological 1-complex. By general position again, K can be made disjoint from such a 1-complex, and then its regular neighborhood, by a topological ambient isotopy. In either case the homeomorphism is smooth near the modification of K, as required for statement (2). ∎

8.2 Smoothing open 4-manifolds

The uniqueness result implies an existence theorem for smooth structures on open manifolds. The most precise version (a classification up to concordance) is obtained using immersion theory, in 8.6B. Here we give a direct proof of existence which avoids the abstract machinery.

Theorem. *A 4-manifold has a smooth structure in the complement of any closed set with at least one point in each compact component. In particular a connected noncompact manifold is smoothable.*

In 8.8 some refinements are given which are more specific about the smooth structure near the singular points.

Proof: First we observe it is sufficient to show that there is a smooth structure on the complement of a discrete set in M. In a connected noncompact 4-manifold there is a locally finite collection of proper flat embeddings of half-open intervals $[0, \infty) \to M$ with a given discrete set as endpoints. A smooth structure on the complement of the set gives a structure on the complement of the arcs, but this complement is homeomorphic to the original manifold. Therefore the noncompact statement follows from the discrete statement. In a compact component a discrete subset is finite, and therefore lies on a flatly embedded closed interval. The complement of such an interval is homeomorphic to the complement of a single point, so we get a smooth structure on the complement of a point. This gives smooth structures on the complement of any nonempty closed set.

The boundary of M is a 3-manifold, so has a smooth structure. Let N_0 be an open collar of the boundary, with the product smooth structure, and let $\theta_i \colon U_i \to M$, for $i \geq 1$ be a locally finite open cover of $M - N_0$ by smooth manifolds (eg. coordinate charts). We inductively construct a discrete set S_i in $N_i = N_0 \cup \bigcup_{j \leq i} U_j$ and a smooth structure on $N_i - S_i$. The construction will have the property that the intersections of S_i and S_{i+1} with $N_i - U_i$ are equal, and the smooth structures agree on $N_i - U_i - S_i$. Consequently the limit yields a smooth structure in the complement of the discrete set $\lim_{i \to \infty} S_i$, proving the theorem.

Suppose that S_i is discrete in N_i, and there is a smooth structure on $N_i - S_i$. Choose a smooth codimension 1 submanifold $V \subset U_i$ so that $\theta(V)$ separates $N_i - \theta(U_i)$ and $\theta(U_i) - N_i$ in $N_i \cup \theta(U_i)$. For example V can be constructed as the inverse of a regular value near 0 of a function $f \colon N_i \cup \theta(U_i) \to (-1, 1)$ so that $f\theta_i$ is smooth, and which takes value 1 on $N_i - \theta(U_i)$ and -1 on $\theta(U_i) - N_i$.

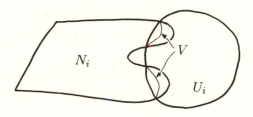

Since S_i is discrete we may assume that $\theta(V)$ is disjoint from it. By theorem 8.1 there is an arbitrarily small isotopy of θ_i to $\tilde{\theta}_i$ which is fixed off $\theta^{-1}(N_i)$, so that $\tilde{\theta}_i$ is smooth near the 2-skeleton of a small displacement of V. Specifically there is a small isotopy of V to $\tilde{V}$ with $\tilde{\theta}_i$ smooth on a neighborhood of $\tilde{V}^2$.

A neighborhood of the 2-skeleton of a 3-manifold is the complement of a union of balls. Therefore there are balls $B_* \subset \tilde{V}$ and an open collar $\tilde{V} \times J \subset U_i$ so that $\tilde{\theta}$ is smooth on $(\tilde{V} - B_*) \times J$. Also, if these isotopies and collars are chosen sufficiently small then $\tilde{\theta}(\tilde{V} \times J)$ will be disjoint from S_i, and will still separate $N_i - \theta(U_i)$ and $\theta(U_i) - N_i$.

Define N'_i to be N_i minus components of $N_i - \tilde{\theta}(\tilde{V} \times J)$ which lie inside $\theta_i(U_i)$, and similarly U'_i to be U_i minus components of $U_i - (\tilde{V} \times J)$ whose images lie inside N'_i. Then $N_i \cup \theta(U_i) = N'_i \cup \tilde{\theta}(U'_i)$, and $N'_i \cap \tilde{\theta}(U'_i) = \tilde{\theta}(\tilde{V} \times J)$. The transition function $\tilde{\theta}$ is smooth on this intersection, except on the balls $\tilde{\theta}(B_* \times J)$. consequently this defines a smooth structure on $N_i \cup \theta(U_i)$ in the complement of $S_i \cap N'_i$ and the balls $\tilde{\theta}(B_* \times J)$. Shrinking the balls to their center points gives a smooth structure on the complement of the discrete set $(S_i \cap N'_i) \cup (\text{centers})$, so defining this to be S_{i+1} completes the induction step. ∎

8.3 Structures in other dimensions

In dimensions less than 4 the categories DIFF, PL, TOP are equivalent. The hardest—and most important to us—is the three-dimensional case:

8.3A Theorem. (E. Moise; Bing [**3**]) *A topological 3-manifold has a PL structure, which is unique up to isotopy.*

By unique we mean that the analog of 8.1 always holds; any homeomorphism is isotopic to a PL isomorphism. The theorem also holds with "PL" replaced by "smooth," as a consequence of the next theorem.

An n-manifold has an n-dimensional tangent bundle, which is smooth, PL or topological depending on the structure of the manifold. (The discussion here will not depend on the details of how "bundles" are defined, but only on formal properties. The microbundles of Milnor [**1**] provide a simple and uniform setting which is satisfactory.) This bundle can be thought of as an infinitesimal version of the structure on the manifold, and therefore provides a way to study the structure.

Suppose M is a manifold with an $\mathcal{M}$ structure ($\mathcal{M}$= DIFF, PL, or TOP), then there is a classifying map for the tangent bundle, $\tau^{\mathcal{M}} \colon M \to \mathbf{B}\mathcal{M}(n)$. Here we are using $\mathbf{B}\mathcal{M}(n)$ to denote the classifying space of the monoid of $\mathcal{M}$ isomorphisms of $\mathbf{R}^n$ fixing the origin. We will use only simple formal properties of this construction; details may be found in Kirby-Siebenmann [**1**, V]. We also note that the inclusion of the orthogonal group $\mathrm{O}(n) \subset \mathrm{DIFF}(n)$ is a homotopy equivalence. This means $\mathbf{B}\mathrm{DIFF}(n)$ may be replaced by $\mathbf{B}\mathrm{O}(n)$, and smooth bundles are equivalent to vector bundles.

If $\mathcal{M} \subset \mathcal{N}$ is one of the natural inclusions of categories then there is a map of classifying spaces $\mathbf{B}\mathcal{M}(n) \to \mathbf{B}\mathcal{N}(n)$. (DIFF is not a subcategory of PL, so technically they are compared via the piecewise differentiable category, which is equivalent to PL in all dimensions. We will disregard this complication here, and pretend DIFF⊂PL.) If M is a manifold with an $\mathcal{M}$ structure, the classifying map for the $\mathcal{N}$ tangent bundle is obtained (up to homotopy) by composing this with the $\mathcal{M}$ tangent bundle map;

$$
\begin{array}{ccc}
M & \xrightarrow{\tau^{\mathcal{M}}} & \mathbf{B}\mathcal{M}(n) \\
\Big\downarrow{\scriptstyle =} & & \Big\downarrow \\
M & \xrightarrow{\tau^{\mathcal{N}}} & \mathbf{B}\mathcal{N}(n)
\end{array}
$$

There is a fibration $\mathcal{M}(n)/\mathcal{N}(n) \to \mathbf{B}\mathcal{M}(n) \to \mathbf{B}\mathcal{N}(n)$. Therefore given an $\mathcal{N}$ bundle there are obstructions in $H^{i+1}(M; \pi_i \mathcal{M}(n)/\mathcal{N}(n))$

to the existence of a refinement to a $\mathcal{M}$ structure. If a refinement to a $\mathcal{M}$ structure is already given on an open set U, then this structure defines a lift there and obstructions to extending it lie in the relative cohomology groups $H^{i+1}(M, U; \pi_i(*))$.

There is also a stabilization $\mathbf{B}\mathcal{M}(n) \to \mathbf{B}\mathcal{M}(n+1)$ obtained by adding a trivial bundle, and the limit of these is denoted by $\mathbf{B}\mathcal{M}$. The composition of the tangent map of M with a single stabilization is the tangent map of $M \times \mathbf{R}$. Therefore existence of a stable factorization of $M \to \mathbf{B}\mathcal{N}$ through $\mathbf{B}\mathcal{M}$ is a necessary condition for an $\mathcal{M}$ structure on $M \times \mathbf{R}^k$, for some k. In fact it is easy to show that this is sufficient as well as necessary (see Milnor [**2**]).

The next theorem asserts that the "infinitesimal" information in the bundle can be "integrated" to solve the manifold problem, except in one case.

8.3B Theorem. *Suppose $\mathcal{M} \subset \mathcal{N}$ is one of the natural inclusions, and if $\mathcal{N} = TOP$ then $n \geq 5$. If M is an $\mathcal{N}$-n-manifold with an $\mathcal{M}$ structure on its boundary then isotopy classes of $\mathcal{M}$ structures extending the given one correspond bijectively to homotopy classes of liftings (rel boundary) of the stable tangent bundle to $\mathbf{B}\mathcal{M}$. Further, $PL/DIFF$ is 6-connected, its homotopy groups are finite, and TOP/PL is an Eilenberg-MacLane space $K(\mathbf{Z}/2, 3)$.*

The PL/DIFF results imply that these two categories are the same up to isotopy, for manifolds of dimension less than 6. Since dimension 3 is covered in 8.3A, the *only* case not completely understood in these terms is the topic of this chapter; smooth structures on topological 4-manifolds.

The PL/DIFF case is a consequence of the "Cairns-Hirsh" theorem, see Hirsh and Mazur [**1**]. The TOP/$\mathcal{M}$ case in dimensions at least 5 are due to Kirby and Siebenmann [**1**]. These can also be proved using the controlled h-cobordism theorem as in 8.1. Homotopy groups of the classifying spaces are calculated by classifying manifolds homotopy equivalent to spheres, see eg. Browder [**1**].

This result uses only the stable tangent bundle. Essentially this implies there is no difference between the stable and unstable bundles. Precisely, we have:

8.3C Theorem. *Suppose $\mathcal{M} \subset \mathcal{N}$ is one of the natural inclusions $DIFF \subset PL \subset TOP$, and if $\mathcal{N} = TOP$ then $n \geq 5$. Then stabilization $\mathcal{M}(n)/\mathcal{N}(n) \to \mathcal{M}/\mathcal{N}$ is $(n + 1)$-connected.*

We recall a map is $(n + 1)$-connected if the relative homotopy groups vanish up to and including dimension $n + 1$. This result will be extended to TOP in dimension 4 in 8.6. There is a rather formal procedure ("immersion theory") which classifies structures up to "sliced concordance" on *open* manifolds, with no category or dimension restriction, in terms of liftings of the *unstable* tangent bundle (see eg. Kirby-Siebenmann [1, essay V]). Theorem 8.3B is a powerful extension of this in almost all ways: it includes compact manifolds, sharpens concordance to isotopy, and includes the stability theorem. However it is weaker in one way; it does not apply to 4-dimensional topological manifolds.

The bundle theory does give invariants in dimension 4, and we summarize this in the next statement. However their vanishing no longer implies the geometric conclusion; the next section shows how far these obstructions are from being sufficient.

8.3D Corollary. *If M is a 4-manifold there is an invariant $\mathrm{ks}(M) \in H^4(M, \partial M; \mathbf{Z}/2)$ which vanishes if M has a smooth (or PL) structure. If $h \colon M \to N$ is a homeomorphism of smooth 4-manifolds there is an invariant $\mathrm{ks}(h) \in H^3(M, \partial M; \mathbf{Z}/2)$ which vanishes if h is isotopic to a diffeomorphism.*

Some properties of this invariant are described in 10.2B.

8.4 Differential geometric results

The study of solutions to certain geometric differential equations on smooth 4-manifolds has provided remarkable insights into the smooth structures themselves. Before stating the results we roughly outline the approach.

A "bundle" over a manifold is a projection which is locally trivial in an appropriate sense. A *connection* on a bundle is essentially a continuous choice of local trivializations at each point. These can be integrated along a path in the manifold to obtain a trivialization of the bundle over the path. In particular it specifies an isomorphism between the fibers of the bundle over the beginning and end points. If this isomorphism is independent of the choice of path between the points, then this process defines a trivialization of the entire bundle.

The *curvature* of a connection gives a measure of the dependence of these isomorphisms on the choice of path. It is a function of 2-planes in the tangent space: roughly one projects the plane into the manifold, use the connection to trivialize the bundle over a small closed loop in the plane, and consider the deviation of the resulting automorphism of the fiber from the identity. Nontriviality of the curvature gives a measure of the nontriviality of the bundle—or at least the failure of the particular connection to give a trivialization.

The special feature of dimension 4 is that the subspace perpendicular to a 2-plane is again a 2-plane. This makes it possible to consider a weaker property than vanishing of the curvature; one can ask if the curvature on a 2-plane is equal to the curvature on the perpendicular 2-plane (or the negative of this). Such connections are called *self-dual* (respectively *anti*-self-dual). The smooth results come from the study of the spaces of these special connections on various bundles over 4-manifolds. Note that it is spaces of connections, rather than any particular connection, which contains the information

For further information see Lawson [1], Freed and Uhlenbeck [1], and the papers of Donaldson, Fintushel and Stern, and Friedman and Morgan listed in the bibliography.

The results give conditions on the quadratic form on the middle dimensional homology $H_2(M; \mathbf{Z})$ of a smooth 4-manifold. Forms, and in particular the invariants "rank" and "signature," and the properties "definite" and "even" are discussed in Chapter 10. We will see there that every form occurs as the form of a topological manifold. In contrast, the smooth result is:

8.4A Theorem. *Suppose M is a closed orientable smooth 4-manifold, with quadratic form λ.*

(1) (Donaldson [1], [3]) *If λ is definite, (ie. rank = |signature|) then $\lambda = \pm kI$.*

(2) (Donaldson [2]) *If M is simply-connected, the form is even, and* (rank $-$ |signature|) ≤ 4, *then $\lambda = \pm k\begin{bmatrix} 0 & 1 \\ 1 & 0 \end{bmatrix}$ (with $0 \leq k \leq 2$).*

This comes close to completely describing the forms which can be realized by closed 1-connected smooth manifolds. The form $jI \oplus k(-I)$ is realized by the smooth manifolds $jCP^2 \# k(-CP^2)$, so this gives the definite forms allowed by (1) and indefinite forms which are not even. An even indefinite form is isomorphic to $jE_8 \oplus k\begin{bmatrix} 0 & 1 \\ 1 & 0 \end{bmatrix}$. The signature 0 ones have $j = 0$, and are realized by $kS^2 \times S^2$. The high dimensional smoothing obstruction in $H^4(M; \mathbf{Z}/2) = \mathbf{Z}/2$ is equal to j mod 2, so j must be even. The Kummer surface Z_4 (see 10.2A) has form $-2E_8 \oplus 3\begin{bmatrix} 0 & 1 \\ 1 & 0 \end{bmatrix}$ with (rank $-$ |signature|) $= 6$, so the inequality in (2) is best possible. It is not yet known what the minimum value of k is for which $2jE_8 \oplus k\begin{bmatrix} 0 & 1 \\ 1 & 0 \end{bmatrix}$ can be realized, when $|j| \geq 2$.

In chapter 10 we will see that a direct sum decomposition of the form corresponds to a topological connected sum decomposition of the manifold. In contrast, the smooth result is:

8.4B Theorem. (Donaldson [2]) *Suppose M is a simply-connected complex algebraic surface with its natural orientation, and is diffeomor-*

phic to a connected sum $N_1 \# N_2$. Then one of N_1, N_2 has negative definite intersection form.

The algebraic process of "blowing up" changes a surface by connected sum with $-CP^2$, so connected sums in which one component has form $-kI$ do occur. The quintic surface Z_5 (locus of $\Sigma_{i=1}^4 z_i^5 = 0$ in CP^3) has form $9I \oplus 44(-I) = 9\left[\begin{smallmatrix} 0 & 1 \\ 1 & 0 \end{smallmatrix}\right] \oplus 35(-I)$, which has many decompositions in which neither component is negative definite. None of these can be realized by smooth connected sum decompositions of Z_5.

Next some restrictions on diffeomorphisms;

8.4C Theorem.

(1) (Donaldson [4], Friedman and Morgan [1], [2]) *There are infinitely many smoothly distinct simply connected complex surfaces ("Dolgachev" surfaces) with form $I \oplus 9(-I)$,*

(2) (Donaldson [4]) *if M is smooth with form $I \oplus k(-I)$, with $k \geq 10$ then there are automorphisms of the form which are not induced by diffeomorphisms of M, and*

(3) (Taubes [1]) *there is a family of smooth structures on $\mathbf{R}^4$ indexed by $\mathbf{R}^2$ which are all smoothly distinct.*

In chapter 10 it is shown that all the manifolds in (1) are homeomorphic to $CP^2 \# 9(-CP^2)$. Since $H^3 = 0$ the bundle theory obstruction cannot detect the fact that these homeomorphisms are not isotopic to diffeomorphisms. Similarly, all automorphisms of the form are induced by homeomorphisms, but according to (2) some of these cannot be smoothed.

We briefly discuss smooth structures on $\mathbf{R}^4$, expanding on statement (3). In every other dimension smooth structures are distinguished by the corresponding smooth structure on the tangent bundle, so a manifold with finitely generated homology has only finitely many structures, and a contractible manifold has a unique structure. In particular in every other dimension every smooth structure on $\mathbf{R}^n$ is isotopic to the standard one.

Shortly after Freedman [2] and Donaldson [1] it was observed by several people that these results implied the existence of exotic structures on $\mathbf{R}^4$. Gompf [1], [2] showed there are more than one, then at least countably many. Taubes [1], as quoted in (3) above, showed there are uncountably many. Gompf [3] has found a topology on a quotient of the set of all structures. Freedman and Taylor [1] found a "universally bad" structure which contains all others. Some of the properties of the Freedman-Taylor structure are discussed in section 8.8.

8.5 Comparisons

The main theorems of this chapter can be considered partial versions of 8.3B, in dimension 4. A connected noncompact 4-manifold has $H^4 = 0$, so 8.3D "predicts" existence of a smooth structure, and this is the conclusion of 8.2. However 8.4A shows that extending across the last point in a compact manifold is quite a different problem. Similarly, since a 2-complex has a neighborhood with $H^3 = 0$, 8.3D "predicts" that a homeomorphism should be smoothable in a neighborhood of one. The smoothing theorem 8.1 falls short of showing this, in that 2-cells of the complex must be modified. However the next statement implies this theorem is sharp, and cannot be improved in any straightforward way.

Proposition.

(1) *There are homeomorphisms* $h: M \to N$ *of closed 1-connected smooth 4-manifolds, and smooth embeddings of* $K = S^2$ *and* S^3 *in* M *so that the bundle obstruction of* h *on* K *is trivial, but* h *is not homotopic to a map which is a diffeomorphism near* K.

(2) *There are homeomorphisms of* $S^3 \times \mathbf{R}$ *into closed 1-connected smooth 4-manifolds, with trivial bundle obstruction but not homotopic to maps which are diffeomorphisms near any displacement of* S^3.

The conclusion is that 8.1 and 8.2 come as close as possible to the high-dimensional versions without contradicting the differential geometric results. We will see in the rest of the chapter that several weakened versions of the problem also follow the high-dimensional pattern. The 4-dimensional pecularities are therefore quite localized.

Proof: Let h be a homeomorphism between $9CP^2 \# 44(-CP^2)$ and the quintic surface Z_5 described after 8.4B (we will see in chapter 10 that such a homeomorphism exists). Let K be the 2-sphere CP^1 in one of the CP^2 summands, or the 3-sphere corresponding to a connected sum decomposition forbidden by 8.4B. If h is homotopic to a map which is a diffeomorphism near either of these then Z_5 is smoothly a connected sum, with CP^2 in the first case, and along the image of K in the second case. In either case this decomposition violates 8.4B.

For the second part of the statement, consider a topological decomposition of the Kummer surface Z_4 as $M \# S^2 \times S^2$. The 3-sphere along which the connected sum is formed gives a homeomorphism of $S^3 \times \mathbf{R}$ into Z_4. Suppose this is homotopic to a map which is a diffeomorphism on a neighborhood V of a displacement. The image of V must separate Z_4, so write Z_4 as a union $K_1 \cup K_2$ of open sets which intersect in the image of V. Similarly, considering V as a subset of S^4 we can write S^4 as a union $L_1 \cup L_2$ with intersection V. The unions $K_1 \cup L_2$ and $K_2 \cup L_1$ are smooth manifolds, and topologically Z_4 is the connected sum $(K_1 \cup L_2) \# (K_2 \cup L_1)$. The forms of the pieces add to give the form of Z_4. Since these forms are nontrivial, one of them violates the conclusion of 8.4A(2). therefore the homeomorphism cannot be diffeomorphism near a displacement of S^3. ∎

8.6 Smoothings of sum-stabilizations

In this section we see that when the manifold can be changed by connected sums with $S^2 \times S^2$, the results predicted by the high-dimensional theory become valid.

Define a *sum-stabilization* of a 4-manifold M to be a connected sum of M with copies of $S^2 \times S^2$ at some discrete subset $T \subset M$. Denote this by $M \# T(S^2 \times S^2)$. Similarly a *sum-stable isotopy* of a homeomorphism is an isotopy of the induced homeomorphism of some sum-stabilizations.

Technically a connected sum $M \# N$ requires choices of embeddings of D^4 in each, and is defined by deleting the interiors of the embeddings and identifying the boundaries via the homeomorphisms with S^3. An embedding $D^4 \to M$ which takes 0 to a point t is determined up to isotopy (keeping 0 at t) by choice of an orientation at t. In the topological category this is a consequence of the (elementary) analogous fact in the smooth category, and the smoothing theorem 8.1(1). In the particular case of sums with $S^2 \times S^2$ the sum is independent of orientations as well, because $S^2 \times S^2$ has an orientation reversing diffeomorphism. Therefore a connected sum of M with copies of $S^2 \times S^2$, one at each point of a discrete subset T, is well-defined up to isomorphism by T. This justifies the notation $M \# T(S^2 \times S^2)$.

If $h \colon M \to N$ is a homeomorphism and $T \subset M$ is a discrete set, then images of disks in M about T are disks in N about $h(T)$. The "induced" homeomorphism $h \# id \colon M \# T(S^2 \times S^2) \to N \# h(T)(S^2 \times S^2)$ is defined to be h on the complement of these disks, and the identity on the copies of $S^2 \times S^2$.

We will take advantage of the structure of the classifying space to be somewhat casual about bundles on sum-stabilizations. Let p denote the projection $M \# T(S^2 \times S^2) \to M$. The pullback of the stable tangent bundle of M is the stable tangent bundle of the connected sum, because the tangent bundle of $S^2 \times S^2$ is stably trivial. Obstructions to existence and uniqueness of smooth structures on these bundles are natural, and lie in $H^4(-; \mathbf{Z}/2)$, $H^3(-; \mathbf{Z}/2)$ respectively (see 8.3B, C). p induces isomorphism of these groups, so we see that homotopy classes of stable bundle smoothings on M and on the the connected sum are in natural one-to-one correspondence. Therefore we can speak unambiguously of a smooth structure on the connected sum "corresponding" to a smooth structure on the stable tangent bundle of M.

Sum-stable smoothing theorem.

(1) *Any smooth structure on the stable tangent bundle of a 4-manifold corresponds to the stabilization of the tangent bundle of a smoothing of a sum-stabilization.*

(2) *Suppose $h: M \to N$ is a homeomorphism of smooth 4-manifolds, with a homotopy of the stable tangent bundle map to a smooth bundle map. Then there is a concordance, and if M is compact and 1-connected there is an isotopy, of a sum-stabilization to a diffeomorphism, so that the induced stable bundle homotopy is homotopic to the given one.*

A concordance is a homeomorphism $M \times I \to N \times I$, which preserves the 0 and 1 levels. An isotopy is stronger in that it commutes with the projections to I. Since the tangent bundle of $M \times I$ is stably the bundle of M, product with I, a concordance defines a homotopy of stable bundle maps between the tangent maps on the two ends. Part (2) asserts that an arbitrary bundle homotopy can be realized geometrically in this way, after sum-stabilization.

There is also a relative version of this, in which a smooth structure is given over an open set, and extensions of the bundle smoothing correspond to extensions of the smooth structure, after stabilization in the complement of the given set.

The realization of smooth structures on *unstable* tangent bundles by sum-stabilizations was proved by Cappell, Lashoff, and Shaneson [1]. This result can be thought of as a compact version of immersion theory, and predates the stability theorem for bundles. The isotopy part of the theorem comes from Quinn [8].

Proof: We will modify the proofs of 8.1 and 8.2 to prove (1) and the concordance part of (2). The refinement of concordance to isotopy in the compact 1-connected case is done in Quinn [8]. It requires an investigation of "pseudoisotopies," and will not be reproduced here.

Suppose a DIFF bundle structure is given on the stable tangent bundle of M. Since $M \times \mathbf{R}$ is 5-dimensional, the high dimensional theorem 8.3B applies to show there is a homeomorphism $h: W \to M \times \mathbf{R}$ with W smooth, and the DIFF bundle τ_W stably equivalent to the given DIFF structure. The first part of the proof is to realize as much as possible of this structure as a product with a structure on M.

We will find a locally finite collection of disjoint flat balls $B_i \subset M$, a smooth structure on $M - \cup_i B_i$, and an isotopy of h to h' which is a diffeomorphism over some neighborhood of $(M - \cup_i B_i) \times \{0\}$. These will be found by induction, fitting together pieces as in 8.2.

Since ∂M has a smooth structure (8.3A), let N_0 be the product structure on an open collar of ∂M. Let U_i, for $i \geq 0$ be a locally finite open cover of $M - N_0 \subset M$ such that there is a smooth structure on U_i and the induced DIFF structure on the tangent bundle of $U_i \times \mathbf{R}$ is equivalent to the one on $h^{-1}(U_i \times \mathbf{R}) \subset W$. For example, subsets of contractible coordinate charts will have these properties. Define $N_i = N_0 \cup \cup_{j<i} U_j$. We will inductively find balls $B_{i,*} \subset N_i$ disjoint, flat, and locally finite in N_i, smooth structures on $N_i - \cup_j B_{i,j}$, and isotopies of h_i to h_{i+1} so that $h_0 = h$ and h_{i+1} is smooth over some neighborhood of $(N_i - \cup_j B_{i,j}) \times \{0\} \subset M \times \mathbf{R}$.

These will also satisfy: the intersections of $B_{i,*}$ and $B_{i+1,*}$ with $N_i - U_i$ are equal, as are the smooth structures, and the isotopy of h_i to h_{i+1} is constant off $U_i \times \mathbf{R}$. These properties imply the collections $B_{i,*}$ and isotopies h to h_i converge to a collection of balls, a smooth structure, and an isotopy satisfying the conditions above.

Suppose the data for N_i has been constructed. The bundle hypothesis on U_i, and 8.3B, imply that h_i can be changed over a neighborhood of $U_i \times \{1\}$ to be smooth on a small product neighborhood of this set. Denote by X the set $(N_i \cap U_i) \times [0, 1]$.

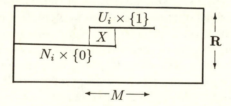

Since h_i is smooth in neighborhoods of both ends of this product, $h_i^{-1}(X)$ is a smooth structure on X agreeing with the given structures on the ends. Projecting to $N_i \cap U_i$ gives a smooth $(\delta, 1)$-connected (δ, h)-cobordism for every $\delta \colon N_i \cap U_i \to (0, \infty)$, as in the proof of 8.1. Choose $\epsilon \colon M \to [0, \infty)$ which is nonzero exactly on $N_i \cap U_i$, and such that for $x \in N_i \cap U_i$, points within $\epsilon(x)$ of x are also in $N_i \cap U_i$. According to 8.1 there is an ϵ isotopy of the given product structure to one which is smooth with respect to the structure on W, except on $Y \subset N_i \cap U_i$. Reversing this isotopy gives an isotopy of $h_i \colon W \to M \times \mathbf{R}$ (rel $M \times \{0\}$, and preserving the 1 level) to h_i' which is a diffeomorphism with respect to the product structure on X, except on Y. Y can be chosen to be an ϵ regular neighborhood of a topological 1-complex.

Choose, as in 8.2, a proper codimension 1 submanifold $V \subset N_i \cap U_i$ which separates $N_i - U_i$ and $U_i - N_i$, in $N_i \cup U_i$, and is disjoint from the balls $B_{i,*}$. Since Y is a neighborhood of a 1-complex we may assume there are 3-balls $A_{i+1,j} \subset V$ so that the intersection of Y with a collar $V \times [-1,1]$ is contained in the collar $\cup_j A_{i+1,j} \times [-1,1]$. (For this make the 1-complex disjoint from the 2-skeleton of V.) Define the collection of balls $\{B_{i+1,j}\}$ to be balls $B_{i,*}$ in the $N_i - U_i$ side of V, together with the $A_{i+1,*} \times [-1,1]$.

Choose a function θ on M which is 0 on the $N_i - U_i$ side of $(N_i \cup U_i) - V \times [-1,1]$, and is 1 on the $U_i - N_i$ side. In $V \times [-1,1]$ require θ to be smooth with respect to the structure on N_i. The graph of θ over $N_i \cup U_i$ is a submanifold of $M \times \mathbf{R}$ which is smooth with respect to the smooth structure h'_i, except on the balls $B_{i+1,*}$. Compose this with the linear second coordinate isotopy of $M \times \mathbf{R}$ which takes the graph to $M \times \{0\}$. This yields h_{i+1} which satisfies the induction hypotheses.

We now work on the missing balls. We have a smooth structure on $M - \cup_i B_i$, so that $h : W \to M \times \mathbf{R}$ is a diffeomorphism over a neighborhood of $(M - \cup_i B_i) \times \{0\}$. Let $r : W \to \mathbf{R}$ denote the composition of h with the projection to $\mathbf{R}$. Smoothness with respect to the product structure implies that the restriction of r to $W - h^{-1}(\cup_i B_i) \times \mathbf{R}$ is smooth near the inverse of 0, and has 0 as a regular value. The next step is to make 0 a regular value of the entire map.

Let $R_i \subset M$ denote disjoint copies of $\mathbf{R}^4$ containing the balls B_i. Change r on a compact neighborhood of $B_i \times \{0\}$ to $\tilde{r}$ which is smooth over some neighborhood of $0 \in \mathbf{R}$, and has 0 as a regular value. The inverse image is unchanged outside some large topological $S^3 \subset R_i$, so topologically $\tilde{r}^{-1}(0)$ is a connected sum $M \# N_i$, with one compact N_i for each ball B_i. (Note however this sphere will usually not be smoothable, so this does not give a smooth decomposition.)

The map $\tilde{r}$ can be changed by homotopy constant outside compact sets in the $R_i \times \mathbf{R}$ so that each inverse image $\tilde{r}^{-1}(0) \cap R_i \times \mathbf{R}$ is connected and simply connected. This is done by "ambient surgery" on embedded 0-, 1-, and 2-handles. This process was developed by Browder [**2**], who provides an elementary description of it. The same technique is also used in the proof of the "end theorem", see eg. Quinn [**2**, §7]. We will not review this here, although it would not be out of place; the end theorem can be modified as in chapter 7 to give a 5-dimensional version which is a useful companion to the h-cobordism theorem. See Quinn [**4**].

This simple connectivity implies that the manifolds N_i in the connected sum decomposition are simply connected. They embed as codimension 1 submanifolds of S^5 (thought of as the 1-point compactification of $R_i \times \mathbf{R}$.) This implies the second Stiefel-Whitney class ω_2 is 0, as is the signature. This identifies the quadratic form on $H_2(N_i; \mathbf{Z})$ as a sum $k\begin{bmatrix} 0 & 1 \\ 1 & 0 \end{bmatrix}$ for some k, which implies that there is a homeomorphism $N_i \simeq k(S^2 \times S^2)$. (There is a simple proof of this in 10.6 using the main embedding theorem; we will not reproduce it here.)

We now have a smooth manifold $\tilde{r}^{-1}(0)$ homeomorphic to $M \# T(S^2 \times S^2)$, for some T. This is a codimension 1 submanifold of W so its smooth tangent bundle stabilizes to give the tangent bundle of W. By construction this is the desired smooth structure on the stable tangent bundle of M, so part (1) of the theorem is complete.

For the concordance part of (2), we suppose $h: M \to N$ is a homeomorphism of smooth manifolds with a stable "smoothing" of the tangent bundle map. Denote the mapping cylinder of h by W, then the high dimensional theorem 8.3B applies to realize the bundle data by a smooth structure on W extending the given one on the boundary. Now consider W as a smooth (δ, h)-cobordism over N. A smooth product structure after sum stabilization is provided by the exercise in 7.2.

As mentioned at the beginning, we do not give a proof of the isotopy part of (2). ∎

8.7 Smoothings of bundles

This section begins with the statement of the stability theorem for 4-dimensional microbundles. Several applications of this to smoothing 4- and 5-manifolds are given, then the bundle theorem is proved using the sum-stability theorem of the previous section.

The main result is an extension of 8.3C to dimension 4;

8.7A Theorem. *The stabilization $TOP(4)/DIFF(4) \to TOP/DIFF$ is 5-connected.*

5-connected means that the relative homotopy groups π_i of the pair $(TOP/DIFF, TOP(4)/DIFF(4))$ are trivial for $i \leq 5$. The vanishing for $i \leq 3$ is from Quinn [4], for $i = 3, 4$ from Lashof and Taylor [1], and for $i = 5$ from Quinn [8].

The immersion theorem described in 8.3 immediately gives:

8.7B Concordance classification. *Concordance classes of smooth structures on a connected noncompact 4-manifold correspond bijectively with homotopy classes of liftings of the stable tangent bundle to $\mathbf{B}DIFF$. These are classified by $[M, \partial M; TOP/DIFF, *] \simeq H^3(M, \partial M; \mathbf{Z}/2)$.*

As a consequence of this we can smooth 4-dimensional submanifolds:

8.7C Submanifold smoothing. *Suppose M is a smooth manifold, N is a 4-dimensional topological manifold without compact components,* and $N \to M$ is a proper locally flat embedding which is smooth near ∂N. Then there is an ambient isotopy of N rel ∂N which takes N to a smooth submanifold.

The most important case of this is the "product structure theorem" where $M = N \times \mathbf{R}$, and the statement is that a smooth structure on $N \times \mathbf{R}$ is isotopic to the product of $\mathbf{R}$ with a smooth structure on N. The theorem is also true for N a 3-manifold, but both the 3- and 4-dimensional statements are false if N is allowed to be compact.

Proof: The first step is to see that N has a normal vector bundle. If the codimension is 1 then this follows from Brown's collaring theorem (see Connelly [1]). In codimension 2 this is given by Kirby-Siebenmann [2].

When the codimension is $k > 2$ then N has a normal structure classified by $\mathbf{B\widetilde{TOP}}(k)$. Here $\widetilde{\mathrm{TOP}}(k)$ denotes the (Δ-set) monoid of block homeomorphisms of S^{k-1}. If by "normal structure" we mean open regular neighborhood, then this comes from Siebenmann [2]. Or, since N is triangulable, we could use topological block bundles as in Quinn [2, 3.3.2 variation 4]. Stabilized microbundles, as in Marin [1] also work. In any case to get a vector bundle it is sufficient to lift the classifying map from $\mathbf{B\widetilde{TOP}}(k)$ to $\mathbf{BO}(k)$.

A lifting is specified on ∂N since it is already smoothly embedded. There are obstructions in $H^j(N, \partial N; \pi_{j-1}\widetilde{\mathrm{TOP}}(k)/O(k))$ to extending this lift over all of N. Since N has no compact components the cohomology vanishes above dimension 3. Therefore for the obstruction group to vanish it is sufficient for $\widetilde{\mathrm{TOP}}(k)/O(k)$ to be 2-connected. This is the case because the stabilization $\widetilde{\mathrm{TOP}}(k)/O(k) \to \mathrm{TOP}/O$ is k-connected, and TOP/O is 2-connected. (The first nontrivial group is $\pi_3 \simeq \mathbf{Z}/2$, which can be identified with the Kirby-Siebenmann invariant.)

Let ν denote the normal bundle of N, then the tangent bundle of M restricted to N is given by $\tau_M|N = \nu \oplus \tau_N$. The stable equation $\tau_N = (\tau_M|N) \oplus \nu^{-1}$ defines a smooth structure on the stable tangent bundle. According to the previous theorem this structure comes from a smooth structure on N. This structure on N induces a smooth structure on the total space of ν, since ν is a vector bundle. This total space also has a smooth structure as an open set in M, and by construction these two smooth structures induce the same smooth structure on the stable tangent bundle. According to the higher-dimensional result 8.3B, there is an isotopy (rel ∂M) which matches up the two structures. This isotopy takes N to a smooth submanifold of M, as required. ∎

As a final application of these methods we give a 5-dimensional analog of 8.2, which gives smooth structures on 4-manifolds in the complement of a discrete set.

8.7D Corollary. *A 5-manifold triad* $(W; \partial_0 W, \partial_1 W)$ *contains a proper 1-dimensional submanifold (with normal bundle)* $T \subset W$ *so that the complement* $(W - T; \partial_0 W - \partial_0 T, \partial_1 W - \partial_1 T)$ *has a smooth structure.*

Proof: First we show it is sufficient to find T such that $\partial_i W - \partial_i T$ has no compact components $(i = 0, 1)$, and the stable tangent bundle $(W - T) \to \mathbf{BTOP}$ lifts to $\mathbf{BO}$. Such a lift restricts to a lift of the tangent bundle of $\partial(W - T)$, so according to the classification theorem 8.7B there is a corresponding smooth structure on $\partial(W - T)$ which extends the unique structure on $\partial_0 W \cap \partial_1 W$. A collar on $\partial(W - T)$ extends this structure to a neighborhood in $W - T$. Now we have a vector bundle structure on the stable tangent bundle of a 5-manifold, which comes from a smooth structure in a neighborhood of the boundary. According to the high-dimensional theorem this implies there is a smooth structure on the manifold.

Now T and the lift are constructed. There is a fibration $\mathbf{BO} \to \mathbf{BTOP} \to \mathbf{BTOP/O}$, so the existence of a lift is equivalent to nullhomotopy of the composition $W - T \to \mathbf{BTOP/O}$.

Up through dimension 5 $\mathbf{BTOP/O}$ is equivalent to an Eilenberg-MacLane space $K(\mathbf{Z}/2, 4)$, so it has a 5-skeleton $S^4 \cup_2 D^5$. Here 2: $S^4 \to S^4$ denotes the third suspension of multiplication by 2, 2: $S^1 \to S^1$. Choose a point in S^4 with a neighborhood over which this map is a 2-fold covering space. Join the preimages in D^5 by a smooth arc. The result is $S^1 \subset S^4 \cup_2 D^5$ with a (nontrivial) normal vector bundle, and such that the complement is contractible in $S^4 \cup_2 D^5$.

Make the map $W \to S^4 \cup_2 D^5$ transverse to this S^1, using the high-dimensional transversality theorem (Kirby-Siebenmann [**1**, III Theorem 1.1]. Denote the preimage by T, then $W - T \to S^4 \cup_2 D^5 \subset \mathbf{BTOP/O}$ is nullhomotopic, as required. The map is easily perturbed so that T intersects each compact component of $\partial_i W$, so the complement has no compact components. ∎

Proof of 8.7A: We want to show that a map from (D^j, S^{j-1}) to $(\text{TOP/DIFF}, \text{TOP}(4)/\text{DIFF}(4))$ deforms into $\text{TOP}(4)/\text{DIFF}(4)$. First suppose $j \leq 4$, then this can be represented by a map from $(\mathbf{R}^j \times \mathbf{R}^{4-j}, (\mathbf{R}^j - B^4) \times \mathbf{R}^{4-j})$. These classifying spaces classify DIFF structures on trivialzed TOP bundles, which we can think of as the tangent bundle of the manifold.

According to the immersion theory result mentioned after 8.3C, there is a smooth manifold and a homeomorphism $M \to (\mathbf{R}^j - B^j) \times \mathbf{R}^{4-j}$ so that the induced smooth structure on the unstable tangent bundle is classified by the given map to $\text{TOP}(4)/\text{DIFF}(4)$. The sum-stabilization theorem 8.6 (the relative version) implies that this smooth manifold extends to $N \to \mathbf{R}^4 \# T(S^2 \times S^2)$ so that the stable smooth tangent bundle of N is the structure classified by the map to DIFF/TOP. Unstably this tangent bundle gives a smooth structure on the tangent bundle of $\mathbf{R}^4 \# T(S^2 \times S^2)$. The difference between this structure and the standard one therefore defines a map $\mathbf{R}^4 \# T(S^2 \times S^2) \to$ DIFF/TOP to DIFF(4)/TOP(4). The composition of this into DIFF/TOP is homotopic to the sum-stabilization of the original map.

If $j \leq 3$ then we can approximate $D^j \to \mathbf{R}^4$ to miss the points T at which the stabilization takes place, and therefore get a deformation of the bundle over D^j (rel S^{j-1}) into DIFF(4)/TOP(4). Therefore the map is 3-connected. When $j = 4$ we get a factorization

$$
\begin{array}{ccc}
D^4 \# T(S^2 \times S^2) & \longrightarrow & \mathrm{DIFF}(4)/\mathrm{TOP}(4) \\
\downarrow & & \downarrow \\
D^4 & \longrightarrow & \mathrm{DIFF}/\mathrm{TOP}.
\end{array}
$$

Obstructions to lifting the map of D^4 (rel S^3) to DIFF(4)/TOP(4) lie in relative cohomology of the left vertical map, with coefficients in the relative homotopy of the right vertical. The cohomology vanishes above dimension 3, while we now know the stabilization map is 3-connected. Therefore all obstruction groups vanish, and the lifting exists. The stabilization is therefore 4-connected.

To see it is 5-connected we use the uniqueness part of the sum-stabilization theorem. Think of a map of D^5 as a homotopy $D^4 \times I \to$ TOP/DIFF from the identity map to itself. Regard this as a homotopy from the tangent map of the identity, to itself. Part (2) of 8.6 asserts this bundle homotopy is induced by an isotopy of a sum-stabilization. An isotopy of 4-manifolds, however, gives a homotopy of the tangent map through 4-dimensional bundle maps. Therefore the homotopy on $(D^4 \# T(S^2 \times S^2)) \times I \to$ TOP/DIFF factors through TOP(4)/DIFF(4). As above the fact that stabilization is already known to be 4-connected implies that this gives a factorization on D^5. Therefore stabilization is 5-connected. ∎

8.8 Almost smoothings

An *almost smoothing* of a 4-manifold is a smooth structure on the complement of a discrete subset $T \subset M$. The subset is called the *singular set* of the almost smoothing. The existence of almost smoothings is asserted by 8.2; here we give some information about what can happen near singularities. First, almost smoothings are obtained whose singular points have particularly nice models. Then "universally bad" singular points are described which overwhelm the smooth phenomena of 8.4.

Two almost smoothings are said to be *diffeomorphic* if there is a homeomorphism which is a diffeomorphism except at the singular points. Similarly we say two singular points are diffeomorphic if they have neighborhoods which are diffeomorphic in this sense.

Examples of almost smoothings of S^4 can be obtained from displacements of the 2-skeleton $S^2 \vee S^2 \subset S^2 \times S^2$. If X is such a displacement (ie. obtained by topological ambient isotopy) then topologically $S^2 \times S^2/X \simeq S^4$, because this is true of the standard embedding and is preserved by ambient isotopy. There is a smooth structure on the complement of the nontrivial identification point in S^4, since this is the same as the complement of X in $S^2 \times S^2$. This defines an almost smoothing of S^4.

8.8A Proposition. (Quinn [6]) *Suppose M is a compact 4-manifold with trivial high-dimensional smoothing obstruction; $\mathrm{ks}(M) = 0$. Then there is an almost smoothing so that each singular point is diffeomorphic to the singular point in an almost smoothing of S^4 obtained by dividing $S^2 \times S^2$ by a displacement of $S^2 \vee S^2$.*

Proof: According to 8.6 there is a smoothing of $M \# T(S^2 \times S^2)$ for some finite $T \subset M$. Apply 8.1 to smooth the inclusion $T(S^2 \times S^2 - \text{ball}) \to M \# T(S^2 \times S^2)$ near a displacement of $T(S^2 \vee S^2)$. M can be recovered by identifying components of this displacement to points. The smooth structure on the sum gives a smoothing in the complement of the identification points, with diffeomorphisms of singular points as required. ∎

Exercise (1) Show that if M is connected and $\mathrm{ks}(M) = 0$ then there is an almost smoothing with singular points diffeomorphic to ones of the form $\pm CP^2/X$, where X is a displacement of $CP^1 \subset CP^2$. (Use the fact that $S^2 \times S^2 \# CP^2 \simeq 2CP^2 \# (-CP^2)$.) (2) Show that if M is connected and $\mathrm{ks}(M) = 0$ then there is an almost smoothing with one singular point, diffeomorphic to the one in some almost smoothing of S^4 with a single singular point. ☐

The first exercise is inspired by the "blowing up" operation of algebraic geometry. This gives maps between complex surfaces, whose nontrivial point inverses are copies of S^2 with neighborhoods diffeomorphic to neighborhoods of $CP^1 \subset -CP^2$. The second exercise makes it clear that the local structure of singular points in almost smoothings gives no information about global smoothings of M.

For the next exercise we note that the smoothing near a singular point gives a smoothing of $\mathbf{R}^4 - 0 \simeq S^3 \times \mathbf{R}$. There is therefore a uniqueness invariant in $H^3(S^3; \mathbf{Z}/2) \simeq \mathbf{Z}/2$ comparing the two structures. If t is a singular point, denote this invariant by $\mathrm{ks}(t)$.

Exercise Suppose M is a compact connected 4-manifold. (3) Suppose an almost smoothing of M is given, with singular set T. Show $\mathrm{ks}(M) = \Sigma_{t \in T}\mathrm{ks}(t)$. (4) Fix a singular point s in some almost smoothing, with $\mathrm{ks}(s) \neq 0$. Show that if $\mathrm{ks}(M) \neq 0$ there is an almost smoothing of M with one singular point diffeomorphic to s and all others as in proposition 8.8A. □

In the opposite direction, we describe some particularly bad singular points. Freedman and Taylor [1] show there is a smooth structure on the half-space $\mathbf{R}^3 \times [0, \infty)$ in which all other smooth structures embed. This structure is determined up to isotopy by this property; denote it by H. The 1-point compactification H^∞ is an almost smooth structure on D^4, with singular set the compactification point $\infty \in S^3$. We say a singular point in an almost smoothing is *universal* if there is an embedding $H^\infty \to M$ which takes ∞ to the singular point and is smooth on the complement.

There are exactly two diffeomorphism classes of universal singularities, distinguished by the invariant ks defined above. Freedman and Taylor show that the complications in the structure of H can be used to absorb smooth embedding problems, so smoothly embedded 2-disks can very often be found. Here we give only the analog of the high-dimensional structure theorem given in 8.4B.

8.8B Proposition. (Freedman-Taylor [1]) *Suppose $f: M \to N$ is a homeomorphism of compact connected almost smooth 4-manifolds. Suppose all singularities are universal, and M, N have the same number of ks $= 0$ points, the same number of ks $= 1$ points, and at least one singular point. Then f is isotopic to a diffeomorphism (of almost smooth structures) if and only if $\mathrm{ks}(f) = 0$.*

CHAPTER 9

Handlebodies, Normal Bundles, and Transversality

The principal conclusions of this chapter are that 5-dimensional topological manifolds have handlebody structures; that submanifolds of 4-manifolds have normal bundles; and that such submanifolds can be made transverse by isotopy. Roughly these mean the basic tools of manifold topology are available for use in 4-manifolds, and we can proceed as in other dimensions.

These results also tie up some loose ends. Transversality implies the topological version of the immersion lemma 1.2. Part I is contingent only on this lemma, so the disk embedding theorems apply to topological manifolds. The technical forms of the h-cobordism theorems (7.1D, 7.2C) require both the immersion lemma and handlebody structures. Therefore the chapter also makes available the h-cobordism theorems for topological manifolds. Finally we note that in each topic treated the results presented here finish the problem; the analogs in other dimensions were settled earlier using other techniques.

The handlebody result is proved in 9.1, and handlebodies in other dimensions are described in 9.2. Similarly the normal bundle results are given in 9.3, and the general situation discussed in 9.4. Normal bundles are obtained as a special case of a powerful flattening theorem. In this context even the assertion that a point has a normal bundle is nontrivial; we apply it to show homology 3-spheres bound contractible 4-manifolds.

The transversality theorem is proved in 9.4, and applied to the immersion lemma in 9.4C. A discussion of transversality in other categories and dimensions is included in 9.5.

9.1 Handlebodies

Suppose W is a manifold, and $N \subset \partial W$ is a codimension 0 submanifold of its boundary. A *relative handlebody structure on* (W, N) is a filtration $W_0 \subset W_1 \subset \cdots \subset W$ so that

(1) W_0 is a collar $N \times I$,
(2) W_{i+1} is obtained from W_i by attaching a handle to $\partial W_i - N$, and

(3) the handles are locally finite; each $x \in W$ has a neighborhood which intersects only finitely many of the sets $W_i - W_{i-1}$.

If the dimension of W is n and $k \leq n$, then W' is obtained from W by attaching a k-handle if $W' = (W \cup D^k \times D^{n-k})/ \sim$, where the equivalence relation $\sim$ is defined by identifying points in $S^{k-1} \times D^{n-k}$ with their images under some embedding in ∂W; see Rourke and Sanderson [**1**, ch. 6], Kirby and Siebenmann [**1**, III.2].

The main result is:

Theorem. *Any 5-dimensional manifold pair W^5, $N \subset \partial W$ has a relative topological handlebody structure.*

Proof: This theorem, and the proof given, come from Quinn [**4**, 2.4]. A slightly weaker result is given by Lashoff and Taylor [**1**]. In outline, we smooth W in the complement of a 1-manifold, and use the smooth structure to get a handlebody structure on this complement. Then we cancel the handles near the 1-manifold, and replace it to get a structure on W.

According to Corollary 8.6D there is a 1-dimensional submanifold $T \subset W$ such that $(W - T, N - \partial T)$ has a smooth structure. The first step is to arrange that T consists of a collection of arcs, with one end on N and the other on $\partial W - N$. This is accomplished by deleting interiors of balls centered at points on T, and adding the boundary to either N or $\partial W - N$. A handlebody structure on this complement gives one on W by considering balls added to N as 0-handles, and balls added to $\partial W - N$ as 5-handles.

These arcs have disjoint neighborhood homeomorphic to $\mathbf{R}^4 \times I$. Deleting these arcs introduces new ends into the manifold, and each of these has neighborhood homeomorphic to $(\mathbf{R}^4 - \{0\}) \times I$, which intersects N in $(\mathbf{R}^4 - \{0\}) \times \{0\}$. Projecting to $\mathbf{R}^4 - \{0\}$ gives this neighborhood the structure of a δ-h-cobordism, any $\delta > 0$. We will apply the technical version of the controlled h-cobordism theorem, 7.2C, to these h-cobordisms. For this let $\mathcal{M} = \text{DIFF}$, $X = \mathbf{R}^4 - \{0\}$, V be empty, and $U = D^4 - \{0\}$, the punctured unit ball. Choose the control function $\epsilon \colon X \to (0, \infty)$ to be $\epsilon(x) = |x|/9$ (the exact form is unimportant as long as it goes to 0 as x goes to 0, and is less than about $1/4$ on $2D^4$).

The local fundamental group of $(\mathbf{R}^4 - \{0\}) \times I$ over X is trivial, so the the K-theory and Whitehead groups of the group ring vanish. Therefore the controlled h-cobordism obstruction group $H_1^{lf}(U^\epsilon; \mathcal{S}(p))$ is trivial, as in 7.2B. Theorem 7.2C implies there is a δ so that if there is a smooth handlebody structure of diameter less than δ on a neighborhood of the inverse image of some large ball, then there is a topological handlebody structure which agrees with the given one outside $U^{2\epsilon}$ and has no handles over U.

Since $(W - T, N - \partial T)$ has a smooth structure, it has a smooth handle-body structure. Choose one, and subdivide it near T so that the projections have diameter less than δ. This provides input to the h-cobordism theorem, so the conclusion is that there is a handlebody structure on $(W - T, N - \partial T)$ which has no handles near T. We claim that replacing T gives a handlebody structure on (W, N).

To see how to replace T, note that $W \simeq (W/ \sim) \cup (\partial W \times I)$, where $W/ \sim$ is obtained by identifying components of T to points. The handlebody structure on $W - T$ is obtained by adding handles to a collar on $N - \partial T$, so by adding limit points as in the endpoint compactification (3.8) we get a pinched collar of N, with handles added outside some neighborhood of the pinched points. But the union of a collar and a pinched collar is a collar on N, so adding $N \times I$ gives the desired handlebody structure on (W, N). ∎

9.2 Handlebody structures in other dimensions

Smooth manifolds all have handlebody structures, obtained from the flow of the gradient vectorfield of a Morse function. PL manifolds have handlebody structures obtained by taking relative regular neighborhoods of simplices in a triangulation. By contrast not all topological manifolds have handlebody structures, and they are difficult to construct when they do exist.

Manifolds of dimension less than 4 have smooth structures (see 8.3A) and therefore handlebody structures. A 4-manifold with a handlebody structure also has a smooth structure: handles are attached by homeomorphisms of 3-manifolds, and according to the 3-dimensional smoothing theorem 8.3A these are isotopic to diffeomorphisms. It follows that the 4-manifolds without handlebody structures are exactly the nonsmoothable ones. According to Donaldson [1], and chapter 10, there are plenty of these.

Manifolds of dimension 6 and greater are shown to have handlebody structures in Kirby and Siebenmann [1, III.2]. The 5-dimensional case covered by 9.4 therefore completes the solution of the problem:

Theorem. *A manifold pair (W, N) fails to have a topological relative handlebody structure if and only if W is an unsmoothable 4-manifold.*

9.3 Normal bundles

The main theorem asserts the existence and uniqueness of normal vector bundles, for submanifolds of 4-manifolds. The existence part is

a consequence of 9.3A, which includes a homotopy criterion for local
flatness, for the ambient space to be a manifold, and even gives normal bundles for some 3-dimensional homology manifolds which are not
manifolds. Uniqueness is proved in 9.3C.

The flattening theorem for points is used to show (in 9.3B) that 3-
dimensional homolgy spheres bound contractible 4-manifolds. This is
derived here to illustrate the power of the theorem, and for use in the
next section; there is a more routine derivation using surgery in 11.4.

A normal bundle for a subspace $N \subset M$ is a vector bundle $E \to N$
together with an embedding $E \to M$ which on the 0-section of E is
the inclusion of N. To avoid inessential problems with uniqueness we
also require normal bundles to be *extendable*. This means if $E \to F$ is
a radial homeomorphism of E with an open convex disk bundle in the
vector bundle F, then the embedding of E extends to an embedding of
F. For example the identity $E = E$ is a nonextendable normal bundle
for the 0-section in E. A radial homeomorphism with the unit disk
bundle (with respect to some Riemannian metric) gives an extendable
normal bundle, and these two embeddings are not ambient isotopic.

Theorem. *A locally flat submanifold of a 4-manifold has a normal bundle, which is unique up to ambient isotopy.*

The technical versions below are also relative: if a bundle is given
over a neighborhood of a closed subsed $K \subset N$ then there is a normal
bundle which agrees with the given one on a (smaller) neighborhood
of K. Similarly two bundles which agree near K are isotopic rel the
restriction to K.

9.3A Theorem. *Suppose $N \subset M$ is a closed subspace with a normal
bundle over a neighborhood of a closed set $N \supset K \supset \partial N$,*

(1) *M is an ANR homology 4-manifold and $M - N$ is a manifold,*
(2) *$N \times \mathbf{R}$ is a manifold, and*
(3) *the local fundamental groups of the complement of N are trivial
    if $\dim N \neq 2$, and infinite cyclic if $\dim N = 2$.*

*Then there is a normal bundle for N in M which agrees with the given
one on some neighborhood of K.*

We explain the conditions in (1)–(3). A homology n-manifold (without
boundary) is a space which satisfies $H_i(X, X - x; \mathbf{Z}) = \mathbf{Z}$ when $i = n$, and
$= 0$ otherwise, at each point $x \in X$. A manifold is an ANR homology
manifold, so the first hypothesis in (1) follows from the second except
at points in N.

In (2), we note that if $N \times \mathbf{R}$ is a manifold and $\dim N \leq 2$ then N is
a manifold. However there are examples of 3-dimensional nonmanifolds

such that $N \times \mathbf{R}$ is a 4-manifold (Bing [2], Andrews-Rubin [1]), and in fact as far as is known at the time of writing every 3-dimensional ANR homology manifold may have this property (see Cannon [1]). The theorem applies to give normal bundles for these spaces.

The local fundamental group condition in (3) holds for locally flat submanifolds, so the conclusion is that correct local fundamental groups implies correct geometry. If the codimension is not 2 then the condition is formally defined by: given $x \in N$ and a neighborhood U in M there is another neighborhood $V \subset U$ so that a loop in $V - N$ is nullhomotopic in $U - N$. Note that if the codimension is 1 then the complement is not locally connected; the condition requires each local complement component to be 1-connected near N.

If the codimension of N is 2 then the condition is: given U as above there is V so that the image of $\pi_1(V - N) \rightarrow \pi_1(U - N)$ is infinite cyclic. The image must be at least this big because this can be detected homologically. If U is sufficiently small then Lefschetz duality gives a surjective homomorphism $H_1(U - N; \mathbf{Z}) \rightarrow \mathbf{Z}$. In these terms the condition is that there is V so that a loop in $V - N$ is contractible in $U - N$ if its image in $\mathbf{Z}$ is trivial.

Finally we observe that a collaring theorem for boundaries of homology manifolds can be obtained from this by applying the theorem to the double along the boundary.

Proof: The proof given basically comes from Quinn [4], and deduces the result from the analogous theorem in dimension 5 and the controlled h-cobordism theorem. The higher dimensional (≥ 5) version as stated comes from Quinn [2, 3.4]. With the additional asumptions that M and N are manifolds, it was proved by Bryant and Seebeck [1] in codimension ≥ 3, Chapman [2] and Seebeck (unpublished) in codimension 2, and Ferry [1], and Cernavskii, and Seebeck (unpublished) in codimension 1. In codimension 2 the high-dimensional result requires the stronger hypothesis of "locally homotopically unknottedness." We show in lemma 9.3B that in dimension 4 this is implied by the fundamental group condition.

According to the higher-dimensional result (using 9.3B when the codimension is 2), $N \times \mathbf{R}$ has a normal bundle in $M \times \mathbf{R}$. A vector bundle over $N \times \mathbf{R}$ splits as a product of $\mathbf{R}$ and a bundle over N; let $E \rightarrow N$ denote this bundle, and let $f : E \times \mathbf{R} \rightarrow M \times \mathbf{R}$ denote the embedding. Choose a Riemannian metric on E, and let $1E$ denote the unit disk bundle. Define W to be $\big(M \times [0, \infty)\big) \cap \big(f(1E \times (-\infty, 1])\big)$.

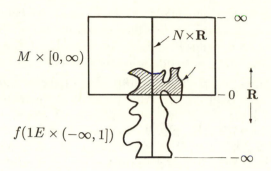

Define $p\colon W \to N \times [0, \infty)$ to be given by the norm in each fiber of E. Note that $p^{-1}(N \times \{0\}) = N \times [0, 1]$. We claim that given $\delta\colon N \times (0, \infty) \to (0, \infty)$ there is a self-homeomorphism h of $N \times [0, \infty)$ which commutes with projection to N such that $hp\colon (W - N \times [0, 1]) \to N \times (0, \infty)$ is a δ-h-cobordism over $N \times (0, 1)$.

The theorem follows from this claim: this is an h-cobordism between neighborhoods of N in $f((1E - N) \times \{1\})$ and $(M - N) \times \{0\}$. The local fundamental group is that of S^{k-1}, where k is the codimension of N, so it is locally constant and either 0 or $\mathbf{Z}$. In either case it is good, and the h-cobordism obstruction group is trivial. Choose $\epsilon\colon N \times (0, \infty) \to (0, \infty)$ to be 1/10 times projection on the second factor. The controlled h-cobordism theorem 7.2C implies that we can find h so that the h-cobordism has an ϵ product structure over $N \times (0, 9/10)$. In particular this gives a homeomorphism of the ends of the h-cobordism, which by choice of ϵ extends by the identity on N. Since any neighborhood of the 0-section of E contains a copy of E, this gives an embedding $E \to M$ which is the identity on the 0-section. This is the required normal bundle.

A little more is required for the application of 7.2C: this applies to h-cobordisms in a category of manifolds which satisfies the immersion lemma, and the immersion lemma in the topological category will not be established until 9.5C. Therefore we show that a neighborhood of the 0-section in W can be smoothed, and apply the theorem in the category $\mathcal{M} = \mathrm{DIFF}$. The top of W is an open 4-manifold, so has a smooth structure by 8.2. The proof below shows a neighborhood of the 0-section deformation retracts in W into this, so the smooth structure on the stable tangent bundle of the top extends to a smooth structure on the stable tangent bundle of the neighborhood. Since the neighborhood is noncompact, the high-dimensional structure theorem 8.3B implies there is a corresponding smooth structure on the neighborhood.

We now verify the claim that W can be made a δ-h-cobordism. This is rather technical, and can safely be skipped.

The first step is to define deformation retractions of W near $N \times [0,1]$ to the two ends. $M \times [0,\infty)$ deforms to $M \times \{0\}$ (push in the $[0,\infty)$ coordinate, straight down in the picture above). Restricting to W gives a deformation of W in $M \times [0,\infty)$ to one end, but this does not keep W in itself. There is also a deformation of the image $f(E \times \mathbf{R})$ to $f(E \times (-\infty, 1])$ defined by pushing down in the $\mathbf{R}$ coordinate. Restrict the first deformation to a neighborhood of $N \times [0,1]$ which stays in the image of f, then use the second deformation to push the track of the first into W. This gives a deformation of a neighborhood of $N \times [0,1]$ in W to one end. Similarly pushing up in the two coordinate systems can be used to define a deformation to the other end.

The situation is now that there is a neighborhood W_0 of $N \times [0,1]$ in W, with deformation retractions $R_0, R_1 \colon W_0 \times I \to W$ in W to $\partial_0 W$ ($= W \cap M \times \{0\}$), and to $\partial_1 W$ ($= W \cap f(E \times \{1\})$). Further, these deformations preserve the complement of $N \times [0,1]$, and on $N \times [0,1]$ are the standard deformations to the ends. Finally, the control map $p \colon W \to N \times [0,\infty)$ is proper, the compositions of the deformations with this are proper, and $p^{-1}(N \times \{0\}) = N \times [0,1]$.

At this point we restrict to a compact situation, to avoid some of the idiosyncracies of estimation with control functions. Specifically we show that if $K \subset N$ is compact then there is a homeomorphism $h \colon [0,\infty) \to [0,\infty)$ so that $(1 \times h)p \colon W - (N \times [0,1]) \to N \times (0,\infty)$ is a δ-h-cobordism over $K \times (0,1)$. To get the general result from this choose an increasing sequence of compacts sets K_i with $\bigcup K_i = N$. Then either use h_i on each K_i to construct a global h, or use the relative form of the controlled h-cobordism theorem to extend product structures over K_i to structures over K_{i+1}, and inductively over all of N.

The compactness is used to eliminate the N coordinate. Denote by d_N the metric on N, and use the metric on $N \times [0,\infty)$ given by $d((x,r),(y,s)) = \sup\{d_N(x,y), |r-s|\}$. Denote the N and $(0,\infty)$ coordinates of the control function p by p_1 and p_2 respectively. Then there are functions $a, b, c, d \colon (0,\infty) \to (0,\infty)$ continuous monotone with limit 0 at 0, such that

(1) $d_N(p_1(R_i(x,s)), p_1(x)) \le a(p_2(x))$, if $p_1(x) \in K$
(2) $b(p_2(x)) \le p_2(R_i(x,s)) \le c(p_2(x))$, if $p_1(x) \in K$, and
(3) $d(t) \le \delta(y,t)$ for $y \in K$.

To define $b(t)$, for example, simply begin with $\inf\{p_2(R_i(x,s)) \mid p(x) \in K \times \{t\}\}$. Because K is compact this defines a continuous function of t. It is positive because R preserves $p_2^{-1}((0,\infty))$ (the complement of

$N \times I$). It has the right limit at 0 because R preserves $N \times I$. Bound it below by a monotone function to get b.

In order for $(1 \times h)p$ to be a δ-h-cobordism over $K \times (0, 1)$ we need

$$\sup\{d_N(p_1 R(x, s), p_1(x)), |hp_2 R(x, s) - hp_2(x)|\} < \delta(p_1(x), hp_2(x)),$$

when $(p_1(x), p_2(x)) \in K \times (0, 1)$. Using the estimates above it is sufficient to find h such that

$$\sup\{a(t), |hc(t) - hb(t)|\} < d(h(t)),$$

for $t \in (0, 1)$.

Choose numbers $\gamma_n > 0$ monotone decreasing to 0 such that

(i) $a(\gamma_n) < \frac{1}{n}$,

(ii) $\gamma_n < b(\gamma_{n-1})$, and

(iii) $c(\gamma_n) < \gamma_{n-1}$.

These are obtained by induction; given γ_j, for $j < n$ satisfying these conditions it is possible to choose γ_n, and therefore inductively an entire sequence.

Now define $h(\gamma_*)$ inductively by setting $h(\gamma_0) = 2$ and $h(\gamma_{n+1}) = \max\{d^{-1}(\frac{1}{n}), h(\gamma_n) - \frac{1}{3(n+2)}\}$. These values are strictly decreasing, so this can be extended to a homeomorphism $h\colon [0, \infty) \to [0, \infty)$ by extending linearly between these values. We verify that h satisfies the estimates above, so yields a δ-h-cobordism over K.

Suppose $t \in (0, 1)$, and let n be such that $t \in [\gamma_{n+1}, \gamma_n]$. Then $a(t) < a(\gamma_n) < \frac{1}{n}$, by (i). On the other hand $d(h(t)) > d(h(\gamma_{n+1})) \geq d(d^{-1}(\frac{1}{n})) = \frac{1}{n}$, so $a(t) < d(h(t))$.

For the second inequality note $|hc(t) - hb(t)| \leq hc(\gamma_n) - hb(\gamma_{n+1}) \leq h(\gamma_{n-1}) - h(\gamma_{n+2})$, using (ii) and (iii). In the definition of h it was required that $h(\gamma_{j+1}) > h(\gamma_j) - \frac{1}{3(j+2)}$, so $h(\gamma_{n+2}) > h(\gamma_n) - \frac{1}{3(n+3)} > \cdots > h(\gamma_{n-1}) - \frac{1}{3(n+1)} - \frac{1}{3(n+2)} - \frac{1}{3(n+3)} > h(\gamma_{n-1}) - \frac{1}{n}$. Therefore $h(\gamma_{n-1}) - h(\gamma_{n+2})$ is less than $\frac{1}{n}$. We saw above that $\frac{1}{n} < d(h(t))$, so $|hc(t) - hb(t)| < d(h(t))$, as required ∎

This completes the proof of 9.3A, except for the following lemma:

9.3B Lemma. *Suppose $N^2 \subset M^4$ is a closed embedding such that the local fundamental groups of the complement are infinite cyclic. Then N is locally homotopically unknotted.*

Proof: In this proof we denote the infinite cyclic group by J, to avoid confusion with the coefficient ring $\mathbf{Z}$. Homology groups are understood to have $\mathbf{Z}$ coefficients unless $\mathbf{Z}[J]$ coefficients are specified.

There are three steps; first a reduction showing it is sufficient to find certain sets satisfying a condition on $\mathbf{Z}[J]$ coefficient homology. Then sets are constructed satisfying the analogous condition on $\mathbf{Z}$ coefficient homology. Finally—and this is the step special to dimension 4—we show that the $\mathbf{Z}$ coefficient condition implies the $\mathbf{Z}[J]$ condition.

Suppose U is a neighborhood in M of a point $x \in N$. Local homotopy unknottedness at x is the condition that there is a smaller neighborhood V such that the image of $\pi_1(V - N) \to \pi_1(U - N)$ is infinite cyclic, and any map $S^k \to V - N$ with $k > 1$ is nullhomotopic in $U - N$.

We begin by identifying the fundamental group more invariantly. Suppose U is a closed ball in M, whose intersection with N lies inside a compact orientable submanifold (eg. a ball). Then Lefschetz duality, the coboundary in cohomolgy, and excision give a canonical homomorphism

$$H_1(U - N) \simeq H^3(U, N \cup \partial U) \xleftarrow{\simeq} H^2(N \cap U, N \cap \partial U) \to H^2(N \partial N) = J.$$

Composition with the Hurewucz homomorphism gives a homomorphism $\pi_1(U-N) \to J$. For any subset of $U-N$, homology with $\mathbf{Z}[J]$ coefficients will mean with respect to this particular homomorphism from π_1 to J.

For the first step we show that given U it is sufficient to find 64 smaller neighborhoods of x, $U \supset U_1 \supset \cdots \supset U_{64}$, and a sequence of neighborhoods of N, $W_1 \supset \cdots \supset W_n \supset \cdots$ so that for $1 \le i \le 64$ and all n the following conditions are satisfied:

(1) $\bigcap_n W_n = N$,
(2) $U_i - W_n$ is connected,
(3) the image of $\pi_1(U_i - W_n) \to \pi_1(U_{i-1} - W_{n+1})$ is taken isomorphically to J by the canonical homomorphism, and
(4) for each $j > 0$ the homomorphism $H_j(U_i - W_n; \mathbf{Z}[J]) \to H_j(U_{i-1} - W_{n+1}; \mathbf{Z}[J])$ is trivial.

According to the "eventual Hurewicz theorem," Ferry [**1**, lemma 3.1] the composition of sufficiently many homologically trivial maps is homotopically trivial. Specifically, given n there are 64 maps, $U_{64} - W_n \rightarrow \cdots \rightarrow U_1 - W_{n+63} \rightarrow U - W_{n+64}$ whose J-covers are trivial on π_1 and homology. The conclusion is that the composition $U_{64} - W_n \rightarrow U - W_{n+64}$ is nullhomotopic when composed with any 4-complex $K \rightarrow U_{64} - W_n$. (The proof that 64 is sufficient is explicit in Quinn [**2**, 5.2], but easily obtained from Ferry's argument). Since $U_{64} - W_n$ itself is a 4-complex, this means the map is nullhomotopic.

We verify that $V = U_{64}$ satisfies the conditions in the definition of local homotopy unknottedness (with respect to U). Suppose $S^k \rightarrow V - N$ is a map, which has trivial image in J if $k = 1$. Since S^k is compact the image lies in some $U_{64} - W_n$, for some n. The map lifts into the J-cover, so by the previous paragraph the composition into $U - W_{n+64} \subset U - N$ is nullhomotopic, as required.

For the second step we construct U_i and W_j satisfying somewhat weaker hypotheses. Suppose U is contained in a ball in M, and is a compact smooth submanifold with respect to the smooth structure on the ball. Choose compact smooth manifold neighborhoods U_i of x, $1 \leq i \leq 192$ so that U_{i+1} is contained in the interior of U_i. (We need $192 = 3(64)$ since $2/3$ of them will be lost in the next step.) Using the local fundamental group hypothesis we can arrange the inclusions $U_i - N \subset U_{i-1} - N$ to have fundamental group image J. Further, using Lefschetz duality as in the construction of the homomorphism to J, we can arrange them to be trivial on higher homology with $\mathbf{Z}$ coefficients. Finally, arrange them to be connected by deleting components not containing x.

The sets W_n are to be chosen to be compact smooth submanifolds whose interiors contain $N \cap U$, ∂W_n is transverse to ∂U_i, all n, i, and satisfying

   (i) $\bigcap_n W_n \subset N$,
   (ii) $U_i - W_n$ is connected,
   (iii) the image of $\pi_1(U_i - W_n) \rightarrow \pi_1(U_{i-1} - W_{n+1})$ is taken isomorphically to J by the canonical homomorphism, and
   (iv) for each j the homomorphism $H_j(U_i - W_n; \mathbf{Z}) \rightarrow H_j(U_{i-1} - W_{n+1}; \mathbf{Z})$ is trivial.

Note these are the same as (1)–(4) in step 1, except for the coefficients in (iv).

We begin the construction of the W_n with a technique for making sets of the form $U_i - W$ connected. Suppose this is the case for $i > k$, and consider components of $U_k - W$. Each of these must intersect $U_{k-1} -$

(int$U_k \cup N$), which is connected. Join components by smooth arcs in this set, and form W' by deleteing open smooth tubular neighborhoods of these arcs from W. This does not change the intersections with U_i for $i > k$, and enlarges $U_k - W$ to make it connected. By induction there is $\tilde{W} \subset W$ so that each $U_i - \tilde{W}$ is connected.

Now construct W_0: Begin with some smooth manifold neighborhood of $N \cap U$ with boundary transverse to ∂U_i and disjoint from the image of some $S^1 \to U_{192} - N$ representing a generator of J. Then make complements connected as above.

Proceeding by induction, we presume W_k is defined for $k < n$, and construct W_n. Note the conditions only concern the inclusion $U_i - \text{int}W_{n-1} \to U_{i-1} - W_n$. The set $U_i - \text{int}W_{n-1}$ is a compact smooth manifold, so has finitely presented π_1 and finitely generated homology. Choose finitely many elements of π_1 which normally generate the kernel of $\pi_1(U_i - \text{int}W_{n-1}) \to J$, and finitely many generators for $H_j(U_i - \text{int}W_{n-1}; \mathbf{Z})$, for $j > 1$. These are trivial in $\pi_1(U_{i-1} - N)$ and $H_j(U_{i-1} - N; \mathbf{Z})$ respectively, by choice of U_*. Therefore they are trivial in some compact subset of $U_{i-1} - N$, so there is a neighborhood W_n' of $N \cap U$ disjoint from this compact subset. Choosing W_n' to be a compact manifold, etc. gives a set satisfying all the conditions except the connectedness in (ii). Modify this by the procedure above to obtain W_n satisfying all the conditions.

By induction a sequence $\{W_*\}$ satisfying conditions (i)–(iv) can be constructed, so the second step is complete.

The final step is to show that the sets U_{3*}, W_{3*} constructed in the second step, satisfy the conditions of the first step. Denote three of the inclusions $U_{i+3} - \text{int}W_n \subset \cdots \subset U_i - \text{int}W_{n+3}$ by $P_3 \subset P_2 \subset P_1 \subset P_0$. Then it is sufficient to show $H_j(P_3; \mathbf{Z}[J]) \to H_j(P_0; \mathbf{Z}[J])$ is trivial for $j > 0$.

The first ingredient is to show the restriction $H_1(P_0, \partial P_0; \mathbf{Z}[J]) \to H_1(P_2, \partial P_2; \mathbf{Z}[J])$ is trivial. Let $\hat{P}_*$ denote the covering spaces associated to the homomorphisms $\pi_1(P_*) \to J$, then it is sufficient to show that any two points in $\partial \hat{P}_1$ can be joined by an arc in $\hat{P}_1 - \hat{P}_2$, and this will be done in the next paragraph. Given an arc in $(\hat{P}_0, \partial \hat{P}_0)$ representing an element in $H_1(P_0, \partial P_0; \mathbf{Z}[J])$ consider an entry point of this arc into $\hat{P}_1$, and the next exit point. Construct a new arc by replacing the segment between these points by an arc in the complement of $\hat{P}_2$. These differ by a loop in $\hat{P}_1$, which by condition (iii) is nullhomotopic in $\hat{P}_0$. Therefore the new arc represents the same class in $H_1(P_0, \partial P_0; \mathbf{Z}[J])$ as the original. But the new arc has fewer components of intersection with $\hat{P}_2$. By induction the class can be represented by an arc disjoint from

P_2, so restricts to the trivial element in $H_1(P_2, \partial P_2; \mathbf{Z}[J])$, as required.

The fact that any two points in $\partial \hat{P}_1$ can be joined by an arc in $\hat{P}_1 - \hat{P}_2$ follows from the corresponding fact in the quotients P_*, and the fact that at least one boundary component has connected preimage in $\hat{P}_1$. The boundary component fact follows from the Van Kampen theorem and the assumption made at the beginning of the proof that all these sets are contained in a ball in M. The arc fact follows from the triviality on H_3 in condition (iv). To see this, begin with an arbitrary arc joining the two points, and consider the first and last points of intersection with a component of ∂P_2. Replace the segment between these by an arc lying in ∂P_2. If the signs of the intersection points are opposite then this new arc can be pushed off the component to eliminate all intersections with it. If the signs are the same then the new arc can be arranged to have exactly one point of transverse intersection with the component. But one point means the intersection number between the homology class of the arc and the class of the component, in $H_3(P_1)$ is nontrivial. This cannot occur, since the class of the component lies in the image of $H_3(P_2)$, which is trivial.

We can now deduce that the inclusion $P_2 \to P_0$ induces the trivial homomorphism on $H_k(*; \mathbf{Z}[J])$ for $k \neq 2$, and on $H^3(*; A)$, for any $\mathbf{Z}[J]$-module A. For H_1 this follows from the fundamental group condition (iii). The homomorphism on H_3 is dual to the cohomology restriction $H^1(P_0, \partial P_0; \mathbf{Z}[J]) \to H^1(P_2, \partial P_2; \mathbf{Z}[J])$, which must vanish because the H_1 restriction is trivial. Higher homology is dual to cohomology in dimensions 0 and below, so also must be trivial. Similarly the homomorphism on $H^3(*; A)$ is trivial since it is dual to the $H_1(*; A)$ restriction, which is trivial because the H_0 and H_1 restrictions with $\mathbf{Z}[J]$ coefficients are trivial.

The vanishing on H^3 implies that $H_2(P_2; \mathbf{Z}[J]) \to H_2(P_0; \mathbf{Z}[J])$ factors through a submodule of a free $\mathbf{Z}[J]$ module. To see this we introduce some notation: let C_k^j denote the chain group $C_k(P_j; \mathbf{Z}[J])$, and let $\partial_k^j : C_k^j \to C_{k-1}^j$ denote the boundary homomorphism. The homomorphism $\partial_3^0 : C_3^0 \to \mathrm{im}\partial_3^0$ defines a class $[\partial_3^0] \in H^3(P_0; \mathrm{im}\partial_3^0)$, which according to the above goes to 0 in $H^3(P_2; \mathrm{im}\partial_3^0)$. This means there is a homomorphism θ making the diagram commute

$$
\begin{array}{ccc}
C_3^2 & \xrightarrow{\ i_3\ } & C_3^0 \\
\downarrow{\scriptstyle \partial} & & \downarrow{\scriptstyle \partial} \\
C_2^2 & \xrightarrow{\ \theta\ } & \mathrm{im}\partial_3^0.
\end{array}
$$

Here i_* denotes the chain map induced by the inclusion $i : P_2 \subset P_0$. The difference $i_2 - \theta \colon C_2^2 \to C_2^0$ vanishes on the image of ∂_3^2, and preserves kernels of ∂_2. It therefore defines a homomorphism $(\ker\partial_2^2/\mathrm{im}\partial_3^2) = H_2(P_2; \mathbf{Z}[J]) \to \ker\partial_2^0$. The composition with the quotient $\ker\partial_2^0 \to H_2(P_0; \mathbf{Z}[J])$ is the inclusion. This therefore defines a factorization of H_2 of the inclusion through the module $\ker\partial_2^0 \subset C_2^0$.

We now complete the proof, by showing $H_2(P_3; \mathbf{Z}[J]) \to H_2(P_0; \mathbf{Z}[J])$ is trivial. Let $S^1 \subset P_4$ be an inclusion which goes to a generator of J. Then these homology groups are the same as the relative homology groups $H_2(P_*, S^1; \mathbf{Z}[J])$. The homomorphisms $H_k(P_3, S^1; \mathbf{Z}[J]) \to H_k(P_2, S^1; \mathbf{Z}[J])$ vanish for $k < 2$, so the integer coefficient homomorphism $H_2(P_3; \mathbf{Z}) \to H_2(P_2; \mathbf{Z})$ is obtained from the $\mathbf{Z}[J]$ coefficient one by applying $\otimes_{\mathbf{Z}[J]}\mathbf{Z}$. Therefore condition (iv) is equivalent to the image of $H_2(P_3; \mathbf{Z}[J]) \to H_2(P_2; \mathbf{Z}[J])$ having the property that $I(\text{image}) = \text{image}$, where $I \subset \mathbf{Z}[J]$ denotes the augmentation ideal. A homomorphism from a module with this property to a free module must be trivial, since $\bigcap_n I^n \mathbf{Z}[J] = 0$. Therefore since $H_2(P_2; \mathbf{Z}[J]) \to H_2(P_0; \mathbf{Z}[J])$ factors through a free module, the composition $H_2(P_3; \mathbf{Z}[J]) \to H_2(P_0; \mathbf{Z}[J])$ must be trivial.

This completes the proof of lemma 9.3B. ∎

The flattening theorem in the simplest case (N a point) implies the following:

9.3C Corollary. *A 3-manifold with the homology of S^3 is the boundary of a contractible topological 4-manifold.*

Proof: Let N denote the homology 3-sphere. The first step is to modify $N \times I$ by "surgery" to get a 1-connected manifold M with the same boundary and homology. This is the "plus construction," and will only be sketched here since it is presented in greater generality and detail in 11.1.

Choose generators for $\pi_1 N$. Since $H_1 N = 0$ there are disjoint smoothly embedded surfaces with trivial normal bundles, $S_1, \ldots, S_n \subset N \times I$ each with a single boundary circle, so that the boundaries represent the chosen generators. These surfaces define a framing of the normal bundles of the boundary circles, so give embeddings $S^1 \times D^3 \to N \times I$. Do "surgery" on these by deleting the interiors, and replacing with copies of $D^2 \times S^2$. Denote the new manifold by M_0, then M_0 is 1-connected and has $H_2(M; \mathbf{Z}) \cong \mathbf{Z}^{2n}$. Generators of H_2 are represented by surfaces $S_i \cup D^2 \times \{*\}$ and 2-spheres $\{*\} \times S^2$, and with respect to this basis the intersection form on H_2 is given by $\begin{bmatrix} 0 & 1 \\ 1 & 0 \end{bmatrix}$.

According to the main embedding theorem there are disjoint topologically framed embedded 2-spheres, with 1-connected complement, representing the same homology classes as the surfaces $S_i \cup D^2 \times \{*\}$. Do "surgery" on these, replacing the interior of copies of $D^2 \times S^2$ with copies of $S^1 \times D^3$, to define M. Homologically this undoes the earlier surgery, so gives a manifold with the same homology as $N \times I$. However since the new embeddings had 1-connected complement, this manifold is 1-connected.

Now construct a space W by stringing together countably many of these, and taking the 1-point compactification.

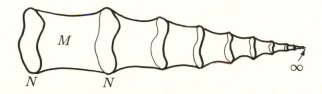

Note that M, and any linear union of copies of it, is homotopy equivalent to S^3. This implies such a union retracts into a single copy, which in turn implies W is contractible. The corollary will be completed by showing W is a manifold. This is the case except possibly at the compactification point, ∞, so we apply 9.3A to this point. Since M is 1-connected, ∞ has locally 1-connected complement. Further, since M is a homology sphere W is a homology manifold at ∞. Finally the proof that it is contractible also shows it is locally contractible at ∞, and for finite dimensional spaces local contractibility is equivalent to being an ANR. Therefore 9.3A applies to show the point has a normal bundle, and W is a manifold. ∎

9.3D Uniqueness of normal bundles. *Suppose $f_i\colon E_i \to M$, for $i = 0, 1$ are normal bundles for N, which agree over a neighborhood of a closed set $K \subset N$. Then there is a bundle isomorphism $b\colon E_1 \simeq E_0$ and an ambient isotopy, rel N and some neighborhood of K, of f_1 to $f_0 b$.*

Proof: As with existence we use a 5-dimensional result and the controlled h-cobordism theorem. Denote the neighborhood of K over which the bundles agree by U. Choose Riemannian metrics for these bundles, which agree over U.

The first step is to use the extendability hypothesis to obtain embeddings in which the originals occur as restrictions to the interiors of the unit disk bundles. The hypotheses are now that the embeddings agree on the disk bundle over U, and the goal is to extend this agreement over the rest of N. Further, since disk bundles of different radii are canonically isotopic, we may restrict to a disk bundle in E_1 of small enough radius that it lies in the image of E_0. Therefore we may replace

M by E_0, and consider E_1 as a second normal bundle for the 0-section $N \subset E_0$.

Next consider the 5-manifold $E_0 \times I$, with submanifold $N \times I$. $E_0 \times \{0\}$ is a normal bundle for $N \times \{0\} \subset E_0 \times \{0\}$, and $E_1 \times \{1\}$ is a normal bundle for $N \times \{1\} \subset E_0 \times \{1\}$. Extend these a short way into the interior of the product by products with $[0, \epsilon)$ and $(1 - \epsilon, 1]$ respectively, then the relative existence theorem for normal bundles in 5-manifolds gives an extension to a normal bundle $F \to E_0 \times I$ for $N \times I$.

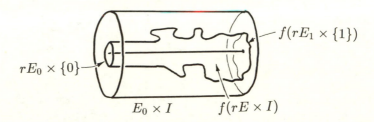

$$rE_0 \times \{0\} \qquad\qquad f(rE_1 \times \{1\})$$
$$E_0 \times I \qquad\qquad f(rE \times I)$$

Since F is a bundle over the product $N \times I$, there is a product structure $F \simeq E_0 \times I$. In particular this gives an isomorphism $E_0 \simeq E_1$. Use this to identify these bundles, and denote them by E, then we have an embedding $f \colon E \times I \to E \times I$ which is the identity on $E \times \{0\}$ and on the restriction $E|U \times I$. The goal is to find an isotopy of f on $E \times \{1\}$ (rel $E|U \times \{1\}$ and $N \times \{1\}$) to an embedding which is the identity on the unit disk bundle.

Suppose $r \colon N \to (0,1)$, and denote the disk bundle of radius r by rE. Suppose r is small enough that the image $f(rE \times I)$ is contained in the unit disk bundle $1E \times I$, and denote the region between these $(= 1E \times I - f((\text{int } rE) \times I))$ by W_r.

The next step is to show that for any $\delta > 0$ there is an r so that W_r is a δ-h-cobordism rel boundary from $(1E - \text{int}(rE)) \times \{0\}$ to $1E \times \{1\} - f(\text{int } rE \times \{1\})$. The control map is the restriction of the projection $p \colon E \times I \to N$, so sizes are measured in N. So we need deformation retractions of W_r to the two ends so the compositions with p have radius less than δ.

We construct the deformation to the $E \times \{0\}$ end; the other differs only in notation. Define $R \colon (E \times I) \times I \to E \times I$ by $R(x, s, t) = (x, st)$, then R is a deformation retraction to $E \times \{0\}$. fRf^{-1} defines a deformation of the image $f(E \times I)$. Since $pfRf^{-1}$ has radius 0 on $N \times I$ there is $r_1 \colon N \to (0,1)$ so that $pfRf^{-1}$ has radius less than δ on $f(r_1 E \times I)$. Choose $1 > s_1 > s_2 > 0$ so that $s_1 E \times I \subset f(r_1 E \times I)$. Finally choose $r > 0$ so that $f(rE \times I) \subset s_2 E \times I$.

We use this data to blend together the retractions R and fRf^{-1}. Choose a function $\phi\colon N \times I \to I$ so that $\phi(y,s) = 0$ if $s > s_1(y)$, and $= 1$ if $s < s_2(y)$. Define $R_1\colon E \times I \times I \to E \times I$ by $R_1(x,s,t) = R(x,s,\max\{t,\phi(x,s)\})$. Since this is the identity on $s_2E \times I$ this induces a deformation of W_r, which is the identity on $W_r \cap E \times \{0\}$. It is the identity for $t = 1$, and for $t = 0$ has image in $W_r \cap (E \times \{0\} \cup s_1E \times I)$. This deformation has radius 0 in N.

Since $s_1 E \times I \subset f(r_1 E \times I)$, the homotopy fRf^{-1} restricted to $f(r_1 E \times I)$ is defined on the image of $R_1(E \times I \times \{0\})$. Composing these two homotopies gives a deformation of W_r to $W_r \cap E \times \{0\}$, and this has radius less than δ by choice of r_1.

The main geometric step is to apply the controlled h-cobordism theorem to W_r. Note the h-cobordism of the boundary is $(\partial 1E) \times I \cup f(\partial rE \times I)$, which has a canonical product structure. Also since f is the identity on $E|U \times I$, W_r has a product structure over U. The local fundamental group of $p\colon W_r \to N$ is $\mathbf{Z}$ if N has codimension 2, and is trivial otherwise, so in any case it is good and the controlled h-cobordism obstruction group is trivial. Therefore Theorem 7.2C implies there is a product structure on W_r extending the one given over some neighborhood of K and on the boundary.

We are using the topological case of theorem 7.2C, which depends on the topological immersion lemma. Although this will not be established until 9.5C, the proof does not depend on the uniqueness of normal bundles, so there is no logical conflict.

This product structure gives in particular a homeomorphism of the ends, $(1E - \operatorname{int} rE) \times \{0\} \simeq 1E \times \{1\} - f(\operatorname{int}(rE \times \{1\}))$, which is the identity on $\partial 1E$ and f on $\partial rE \to f(\partial rE)$. Conjugate the canonical radial isotopy of ∂rE to $\partial 1E$ by this homeomorphism, to obtain an isotopy of f to an embedding which on $\partial rE \times \{1\} \to \partial 1E \times \{1\}$ is multiplication by $\frac{1}{r}$. Finally compose with the radial isotopy of $1E$ to rE.

On the $E \times \{1\}$ end we now have $f\colon E \to E$ which preserves the unit disk bundle, and is the identity on $\partial 1E$, on the 0-section N, and over some neighborhood of $K \subset N$. The proof will be completed by construction of an isotopy of f to an embedding which is the identity on $1E$.

Define a family of automorphisms $A_t \colon E \to E$ by $A_t(x) = \|x\|^{t-1}x$, for $t \in (0, \infty)$. Define $F \colon 1E \times I \to 1E$ by

$$
F(x, t) = \begin{cases} A_{\frac{1}{t}} f_0^{-1} f_1 A_t(x), & \text{for } t > 0 \\ x, & \text{for } t = 0. \end{cases}
$$

Since $f_0^{-1} f_1$ is the identity on $N \cup \partial(1E)$, this is continuous, so defines the required isotopy from $f_1^{-1} f_2$ to the identity. This is a parameterized form of the Alexander isotopy. ∎

9.4 Normal bundles in other dimensions

Every topological submanifold has some sort of normal structure, but the best description of this structure is not yet clear. Vector bundles, and even microbundles, are too rigid. Topological block bundles give the "right theory" and are easy to use, but require a triangulation of the submanifold. Stabilized microbundles (Marin [1]) and approximate fibrations (Quinn [2, §3.3]) both "work," but are not straightforward to work with. See 9.6 for further discussion. However in low dimensions and codimensions vector bundles can be used, and we review this situation.

In codimension 1, the fact that local collaring implies collaring was proved in considerable generality by M. Brown (see Connelly [1]). The existence and uniqueness of normal bundles for locally flat codimension 2 embeddings when the ambient dimension is greater than 4 is due to Kirby and Siebenmann [2].

Low dimensional submanifolds also have normal vector bundles. The map of stable classifying spaces $B_O \to B_{TOP}$ is 3-connected. Therefore a map of a manifold with the homotopy type of a 3-complex, into B_{TOP}, lifts to B_O. This, together with stability theorems and the fact that submanifolds have normal structures stably classified by B_{TOP}, implies that a submanifold of dimension at most 3, in a manifold of dimension at least 5, has a normal vector bundle. This fact is used in the submanifold smoothing theorem, 8.7C.

The cases left unsettled by these results are exactly those covered by 9.3; submanifolds of codimension greater than 1, in a 4-manifold.

The flattening results similarly were known in all other dimensions before the 4-dimensional developments. In higher dimensions the flattening theorem for codimension 2 requires "local homotopy unknottedness," essentially requiring the vanishing of higher local homotopy groups. This is in fact necessary: let $V \subset S^k$ be an embedded S^{k-2} with complementary fundamental group $\mathbf{Z}$, but whose complement does not have the homotopy of S^1. These exist for $k > 4$. Then the cone on (S^k, V) gives a codimension 2 embedding of D^{k-1} in D^{k+1} with infinite cyclic complementary local fundamental groups, but which is not locally flat. The fact that in dimension 4 the fundamental group condition is sufficient, is new here.

9.5 Transversality

The general setting is a pair of proper submanifolds U, $V \to M$ of a 4-manifold, and the objective is to find an isotopy which moves U to a submanifold transverse to V. Here "transverse" means that points of intersection have coordinate neighborhoods which the submanifolds appear as transverse linear subspaces. At boundary points transverse half-spaces are used as models.

There is also a global version of transversality: U has a normal bundle $E \to M$ so that $E \cap V = E|(U \cap V)$. The existence and uniqueness of normal bundles in this dimension implies the local and global versions agree. In higher dimensions the global approach is required; it is necessary to define transversality with respect to a given normal structure.

The boundary of a manifold has only a collar rather than a full normal bundle, but it is reasonable to define "transverse to the boundary" in the same way: V is transverse if there is a collar $\partial M \times I \to M$ so that $V \cap \partial M \times I = (V \cap \partial M) \times I$. With this terminology V is a "proper" submanifold if it is closed, $V \cap \partial M = \partial V$, and V is transverse to ∂M.

9.5A Theorem. *Suppose U, V are proper topological submanifolds of a 4-manifold M, $K \subset U \cap V$ is closed, and U, V are transverse in a neighborhood of K. Then there is an isotopy of M fixed near K and off any given neighborhood of $U \cap V$, so that the final image of U is transverse to V.*

Proof: When U and V have dimension at most 2 this comes from Quinn [4], the other cases are from Quinn [9]. The most difficult case is when U and V both surfaces; lower dimensional cases follow from the smoothing theorems and higher ones by an induction on handles. As usual the difficult case is the most useful, being a principal ingredient of the immersion lemma.

First we eliminate the boundary. The boundary is 3-dimensional, so submanifolds of it can be smoothed and made transverse. Extend the transversality to a neighborhood of the boundary using the collar structures, and add this neighborhood to K. The boundary can now be deleted without changing the problem.

Next, we may assume that M is a smooth manifold and V a smooth submanifold. For this note V has a smooth structure since it has dimension less than 4, and it has a normal vector bundle. The total space of a vector bundle over a smooth manifold has a natural smooth structure. Replace M by this total space, and U by the intersection with the total space.

At this point the argument separates into cases depending on dimensions.

Case 1; dim$U \leq 1$. The transverse set is empty if $\dim U + \dim V \leq 3$, and is discrete if $\dim U + \dim V = 4$. In the latter case $U \cap V - K$ is also closed, so restricting to a neighborhood disjoint from K gives an equivalent situation where again K is empty.

Since $\dim U \leq 1$ the smoothing theorem 8.1(1) implies that U is isotopic to a smooth submanifold. More precisely, there is a isotopy supported in a neighborhood of $U \cap V$ which makes the intersection of U with some smaller neighborhood smooth. The smooth transversality theorem then gives a further isotopy to a submanifold transverse to V.

Case 2; dim$U$ = dim$V$ = 2. The transverse region K is again a discrete set of points, so as above restriction to a neighborhood of $U \cap V - K$ gives an equivalent situation with K empty.

According to theorem 8.1(2), there is a regular homotopy of U supported in a neighborhood of V to an immersion U' whose intersection with a smaller neighborhood is smooth. According to the smooth version of the theorem there is then a isotopy which makes U' transverse to V. The plan is to work back past the singularities in the regular homotopy to obtain an isotopy making U transverse to V.

A regular homotopy is a composition of a sequence of isotopies, finger moves, and Whitney moves. According to the remark after 8.1, only isotopy and a locally finite collection of finger moves are needed to get a smooth, and therefore a transverse U'. However if a Whitney move were encountered it could easily be replaced by an isotopy; reversing direction, U is obtained from a transverse object by a finger move. A finger move takes place in a neighborhood of an arc, which (according to case 1) can be made disjoint from V by an isotopy. A finger move disjoint from V does not change intersections with V, so this isotopy also makes U transverse to V.

After isotopy we can arrange that the finger moves all occur at once, disjointly. Considering them separately, and reversing direction in the regular homotopy, we have reduced the proof to the case where U is obtained from U' by a single Whitney move.

We cannot at this point make V transverse to the Whitney disk; since it is only topologically embedded this would require the transversality result we are trying to prove. However to illustrate the idea of the proof we indicate what to do if it were transverse.

Suppose V is transverse to the Whitney disk. Push V off the disk through U' by finger moves (see 1.5). This gives an isotopy to a surface V' disjoint from the disk and with new transverse intersections with U'. Now use the Whitney isotopy across the disk to move U' back to U. Since this isotopy takes place in a neighborhood of the disk, which is disjoint from V', it does not change the intersections with V'. Therefore U is transverse to V'. The inverse of the finger move isotopy thus gives an isotopy of U to a surface transverse to V, as desired.

Since we cannot directly make V transverse to the Whitney disk, we replace it with a smooth tower and make V transverse to that. It can then be push it off the tower by finger moves. The embedding theorem can be used to recover a topologically embedded disk in a neighborhood of the tower, and this new disk used to recover U from U'.

Denote the Whitney disk by A. Using the lower dimensional case of the theorem we may assume V is disjoint from the boundary of A. The intersection $A \cap V$ is therefore compact. Since it is topologically flatly embedded, there is a neighborhood of the interior of A disjoint from U' and homeomorphic to $\mathbf{R}^4$. Denote this by N. As an open subset of M it has a smooth structure about which we know nothing.

Apply the smoothing theorem once again, to A in a compact neighborhood of the intersection $A \cap V$. This yields a regular homotopy consisting of a finite number of isotopies and finger moves, to a surface A' which is smooth near its intersection with V. Arrange the finger moves to take place all at once, disjointly, and denote by W_i the resulting (disjoint) Whitney disks for the intersections of A'. Again these are only topological, although the part of A where they are attached is smooth.

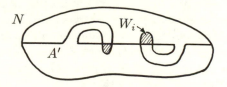

We begin setting up the data to apply the embedding theorem to the W_i. Let X denote the complement of the interior of a closed smooth regular neighborhood of A'. X is a smooth manifold with boundary, namely the boundary of the regular neighborhood. The Whitney disks intersect this boundary in smooth circles.

The next step is to observe that $\pi_1(X)$ is infinite cyclic. X has the same fundamental group as its interior, which is diffeomorphic to $N - A'$. Therefore we show $\pi_1(N - A') \simeq \mathbf{Z}$. $N - A$ is homeomorphic to $\mathbf{R}^4 - \mathbf{R}^2$ so has $\pi_1 \simeq \mathbf{Z}$. Adding the (locally flat) finger arcs to A does not change the fundamental group of the complement. But A union finger arcs has complement homeomorphic with the complement of $A' \cup \bigcup_i W_i$. Finally, adding the Whitney disks back in has the effect of killing the image of the linking circle in the complement. The linking circle bounds the punctured linking torus, so this element is a commutator. Since the fundamental group is cyclic, commutators are already trivial, and the group is not changed by this.

The last part of this argument shows the linking circles of the W_i are nullhomotopic in X. Since N is homeomorphic to $\mathbf{R}^4$, $\omega_2 = 0$ and so there are (smooth) framed immersed transverse spheres for the W_i (see 1.3, corollary 1). The W_i are homotopic to smooth immersions, by the (smooth) immersion lemma. Since the original W_i are embedded, and intersection numbers are homotopy invariants, these smooth immersions satisfy the hypotheses of the corollary of the embedding theorem 5.1.

The proof of the embedding theorem begins by replacing the immersions with capped towers, which are smooth since we are working in a smooth manifold with smooth data. Smooth transversality gives an isotopy which makes V transverse to these towers, i.e. transverse to the component surfaces. Push these intersections down (see 2.5) off the tower to give new transverse intersections of V with A'. Push these off through U'. Reversing this isotopy gives an isotopy which moves the towers and A' disjoint from V, and introduces transverse intersection points between U' and V.

The embedding theorem is completed by finding locally flat disks in a neighborhood of the capped tower. These give new Whitney disks W' for the selfintersections of A'. Using these gives a new embedding A'', which is itself a new Whitney disk for intersections in the immersion U'. A Whitney move across A'' gives a new embedding U''. Since A'' is disjoint from V, such moves do not change intersections with V. Since U' had been arranged to be transverse to V, U'' is also transverse.

The proof will be completed by showing that U is isotopic to U'', which is the effect of the next lemma. For this we need a notation; suppose an open neighborhood of a Whitney disk is parameterized in the standard way as $\mathbf{R}^2 \times \mathbf{R}^2$, with the disk appearing as $D^2 \times 0$ and the base surface inside $S^1 \times \mathbf{R}^2$. Then $\mathrm{int} D^2 \times \mathbf{R}^2$ is a "standard neighborhood of the interior" of the disk.

9.5B Lemma. *Suppose U and U'' are immersed surfaces obtained from*

an immersion U' by Whitney moves across disks A, A'' respectively,
which are equal near U'. If the interior of A'' is contained in a standard
neighborhood of the interior of A, then U'' is isotopic to U.

Proof of the lemma: The isotopy across A takes place in a standard
neighborhood of A, which we denote by N. This neighborhood intersects
U' in two pieces, say U'_1 and U'_2. Denote the isotopy itself by h_t with
$0 \leq t \leq 1$, so h_0 is the identity and h_1 takes U'_1 disjoint from U'_2. The
picture shows a cross-section.

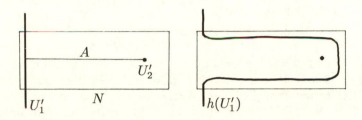

Similarly the move across A'' takes place in N''. Since the interior of A''
is in a standard neighborhood of the interior of A, and $h_1(U'_1)$ is near
the frontier, we may arrange that N'' is disjoint from it. Further, since
the disks agree near the boundary we can arrange that N'' extends out
beyond N near U'_1. This extension can be chosen so that the associated
Whitney isotopy h''_t moves N as well as U'_1 off A' and U'_2.

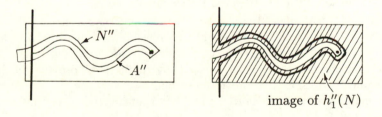

image of $h''_1(N)$

We claim that the Whitney isotopy in the image $h''_1(N)$ gives an isotopy
of U'' to U. Precisely, consider the isotopy $h''_1 h_t (h''_1)^{-1}$, with $0 \leq t \leq 1$.
When $t = 0$ it is the identity, and when $t = 1$ it takes U''_1 to $h''_1(U_1)$.
Since h'' is supported in N'' which is disjoint from U_1, this image is U_1
as desired. Similarly, since $h''_1(N)$ is disjoint from U'_2, this is also left
fixed. The result is therefore an isotopy of U'' to U. ∎

This completes the proof of the lemma, and therefore case 2 of the
theorem.

9.5C Corollary. *The immersion lemma 1.2 is valid in the topological category.*

Proof: The first part of the lemma asserts existence of immersions of ribbons and disks, agreeing up to sign or rotations with ones specified on the boundary. This is now easy to deduce from the smooth version. Suppose $f: N \to M$ is a map of a surface, which restricts to a framed embedding $\partial N \to \partial M$. Let M_0 denote the complement of an appropriate discrete set disjoint from the image of N, so that by theorem 8.2 the result has a smooth structure. According to the smoothing theorem for 3-manifolds, 8.3A, ∂f is isotopic to a smooth framed embedding. (More precisely we should say the smooth structure is isotopic to one in which ∂f is smooth.) Applying the smooth version of the theorem to $f: N \to M_0$ gives immersions in M_0, and thus M.

The second part of the lemma concerns the general position condition built into the definition of immersion we are using (see 1.2). Suppose, as in the lemma, that $f: N \to M$ is an immersion on U and on V, where $U \cup V = N$, and we want to change f by isotopy on U to intersect V transversally in points. Let K be the selfintersection points lying in $U \cap V$; these are already transverse. The selfintersections of $U - V$ are points, so we may arrange them to be disjoint from the image of V. Similarly arrange the selfintersections of $V - U$ to be disjoint from the image of U. Then in a neighborhood of $U \cap V$ both are embedded. Restricting to this neighborhood gives submanifolds to which theorem 9.5A applies. This theorem yields an isotopy fixed off a neighborhood of $(U \cap V) - K$, which makes U transverse to V. The resulting map of $U \cup V$ is the immersion required for the lemma. ∎

Proof of 9.5A, case 3; dim$U = 3$: In situations not covered by the previous cases one of the submanifolds must have dimension 3. Since the statement is symmetric we may assume dim$U = 3$. The dimension of V is either 2 or 3; the proof proceeds by "induction," proving the first case and then using it to prove the second.

The first step is a lemma which is used to get transversality from lower dimensional cases.

9.5D Lemma. *Suppose $M \supset V$ is a proper submanifold of a 4-manifold. Let Y be a manifold of dimension ≤ 3, let $p: F \to Y$ be a vector bundle and $f: F \to M$ a locally flat embedding which on Y is proper and transverse to V, and is transverse to ∂M. Then there is an ambient isotopy of f rel Y to an embedding f' so that f' and its restriction to the sphere bundle of F are transverse to V.*

Proof: Choose a normal bundle for image of the total space $f(F)$ in M, and denote the restriction to Y by F'. The normal bundle can be identified with the pullback p^*F'. The total space of p^*F' is homeomorphic to that of the sum $F \oplus F'$, and the embedding $F \oplus F' \to M$ is a normal bundle for Y.

Since Y is transverse to V there is a normal bundle for Y with good intersection with V. By uniqueness of normal bundles we may assume (after isotopy rel Y) that $F \oplus F'$ has this property, namely $F \oplus F' \cap V = (F \oplus F')|(Y \cap V)$. This implies that F and V are transverse: $p^*F' \cap V = F \oplus F' \cap V = (F \oplus F')|(Y \cap V) = p^*F'|(F \cap V)$.

A similar identity shows transversality of the sphere bundle. Denote this by SF, with projection $sp\colon SF \to Y$, then SF has normal bundle $sp^*F' \oplus \mathbf{R} = F \oplus F' - Y$. The intersection is $(sp^*F' \oplus \mathbf{R}) \cap V = (F \oplus F' - Y) \cap V = (F \oplus F'|V \cap Y) - Y = (sp^*F' \oplus \mathbf{R})|(SF \cap V)$. ∎

The first application of the lemma to the proof of case 3 is to arrange U to have trivial normal bundle. U has codimension 1 so the normal bundle is a line bundle classified by a map $e\colon U \to RP^n$, some n. Since $\dim U = 3$ this map may be made transverse to $RP^{n-1} \subset RP^n$. The submanifold $e^{-1}(RP^{n-1}) \subset M$ is lower dimensional than U, so an earlier case of the transversality theorem applies to give an isotopy making it transverse to V. The lemma makes U transverse to V in a neighborhood of $e^{-1}(RP^{n-1})$. Add a closed subneighborhood to the set K to be held fixed, then $e^{-1}(RP^{n-1})$ can be deleted without changing the problem. The normal bundle of the remainder of U is pulled back from $RP^n - RP^{n-1}$. This is contractible so the normal bundle is trivial.

We also arrange that the complement of a small open neighborhood of K in U in which it is transverse has no compact components. Choose a point in each such component and make it transverse to (disjoint from) V. These points are not in $V \cap U$, so if we restrict future isotopies to lie in a neighborhood of $V \cap U$ not containing them they can be deleted from M without changing the problem.

The next step is to make the embedding of U smooth in a neighborhood of K. A neighborhood of K in $U \cap V$ is a submanifold of V, which is isotopic to a smooth submanifold since $\dim V \leq 3$. Since this submanifold is a transverse intersection, a neighborhood of it in U is given by the restriction of some normal bundle for V. At the beginning of the proof V was arranged to be a smooth submanifold of M, so there is a smooth normal bundle. Using the uniqueness of normal bundles we may assume this is the bundle intersecting U nicely. But the restriction of a smooth bundle to a smooth submanifold is smooth, so this makes U smooth near K.

Use the triviality of the normal bundle of U to find a collar on one side: an embedding $U \times I \to M$ which restricts to the original embedding on $U \times \{0\}$, and which is smooth in a neighborhood of $K \times I$.

The interior of the collar $U \times (0,1)$ has a smooth structure as an open set in M, and this agrees with the product structure in a neighborhood of $K \times I$. Approximate the projection $U \times I \to I$ rel this neighborhood and $U \times \{0,1\}$ by a map g which is smooth on $U \times (0,1)$. Let t be a regular value of g, then $g^{-1}(t)$ is a smooth proper submanifold of M. Arrange by small smooth isotopy rel a neighborhood of $K \times \{t\}$ that it is transverse to V.

Construct W from $g^{-1}([t,1])$ by deleting components disjoint from $U \times \{1\}$. Denote $U \times \{1\}$ by $\partial_1 W$, and define $\partial_0 W$ (a subset of $g^{-1}(t)$) to be the rest of ∂W.

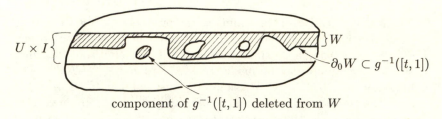

component of $g^{-1}([t,1])$ deleted from W

Next note that W is smoothable. By construction it has a smooth structure on the complement of part of $\partial_1 W$: the complement in $U \times \{1\}$ of a neighborhood of $K \times \{1\}$. But we have arranged that for an appropriate neighborhood of $K \times \{1\}$ this complement has no compact components, so smoothability follows from the submanifold smoothing theorem 8.7C.

Since W is smoothable there is a relative handlebody structure on the pair $(W, \partial_1 W)$. More precisely, since W is already a smooth collar in a neighborhood of $K \times [t,1]$ there is a handlebody structure obtained from a collar on $\partial_1 W$ by attaching handles outside a neighborhood of K.

We now eliminate handles from W. The deletions from $g^{-1}([t,1])$ made in the definition of W make the pair $(W, \partial_1 W)$ connected. Therefore all 0-handles can be cancelled by 1-handles (this is the easiest case of "geometrical connectivity," see Wall [2]).

Let $(D^k, S^{k-1}) \to (W, \partial_0 W)$ be the dual core of a topmost handle in W. Since $\partial_0 W$ is transverse to V the intersection $V \cap (W - \partial_1 W)$ is a proper submanifold. If $k < 3$ then an earlier case of the transversality theorem can be used to make the core transverse to the intersection. Since the handle itself can be considered as the disk bundle of the normal bundle of the core, Lemma 9.5D implies that the boundary of the handle can be made transverse to V by a further isotopy. But this implies that ∂_0 of the handlebody obtained by deleting the handle from W is transverse to V. Replace W by this smaller handlebody.

Since all but the 1-handles of W have duals with core dimension less than 3, this operation can be repeated to obtain a handlebody with only 1-handles and with ∂_0 transverse to V.

Now consider the 1-handles. Since $W \subset \partial_1 W \times I$ the core arcs are homotopic rel ends into $\partial_1 W$. Replace the homotopies by immersed disks, and then use finger moves across the boundary to remove intersections. This gives disjointly embedded 2-disks whose boundaries are the core arcs and arcs on $\partial_1 W$. These 2-disks define 2-handles which cancel the 1-handles. In particular the complement $W' = (U \times I - \operatorname{int} W)$ has a handlebody structure with only 2-handles. (The embedded disks are the cores of the duals of these handles.)

Since the upper boundary of W' is $\partial_0 W$, which is transverse to V, we can repeat the argument above: make the 2-handles transverse to V using an earlier case of the theorem and Lemma 9.5D, and delete. The result is a collar on U with one boundary transverse to V. Therefore U is isotopic to a manifold transverse to V, as required for the theorem. ∎

9.6 Transversality in other categories and dimensions

There are two important settings for transversality; submanifolds, and maps. In the first, two submanifolds of a manifold are given, and the desired conclusion is that one can be moved by isotopy to be transverse to the other. In the map case a map $f: M \to X$ is given, with M a manifold and X a space containing as an open set the total space of an appropriate type of bundle, over a subspace Y. The objective is to change the map by homotopy so that $f^{-1}(Y)$ is a submanifold, whose normal bundle is the pullback of the given bundle.

Map transversality can be deduced from submanifold transversality. For this replace Y by a manifold (by taking a regular neighborhood of an appropriate subcomplex, for example). The total space of the bundle is then a manifold with Y as submanifold. Then change f so that the part of M mapping to the total space is embedded (by crossing Y with some $\mathbf{R}^n$ in which M embeds, for example). Change this embedding of M by isotopy to be transverse to the new Y. Project back to the original Y to obtain a homotopy to a transverse map.

To be more specific we consider the various categories of manifolds separately.

9.6A Smooth manifolds. Vector bundles are used, since all smooth submanifolds have normal vector bundles. Smooth submanifold and (therefore) map transversality hold in all dimensions, and are simple consequences of the Morse-Sard-Brown theorem on critical images. See Guillemin and Pollack [1].

9.6B PL manifolds. In the PL category the appropriate bundles are the "block bundles" developed by Rourke and Sanderson [**1**; **2**]. Submanifold and map transversality hold in all dimensions, and with the correct definitions the proofs are even simpler than the smooth category.

The development in this category had an awkward beginning, using "microbundles." These are easy to define, but normal microbundles for submanifolds do not always exist and are not unique, and this leads inevitably to technical problems. See Kirby and Siebenmann [**1**, III] for a discussion.

9.6C Topological manifolds. The best description of "bundles" in topological category is still not completely settled. Topological microbundles have the same nonexistence drawbacks that PL microbundles do. Marin [**1**] has a variation on microbundles which does have the correct technical properties. If the base of the bundle is triangulable as a simplicial complex then topological block bundles can be used. However not all topological manifolds can be so triangulated. Approximate fibrations are simple and attractive, but proofs of existence and transversality (when known) are so far quite complicated. See eg. Quinn [**2**, section 3.3].

Kirby and Siebenmann [**1**, III] give submanifold and map transversality theorems using topological microbundles, when nothing of dimension 4 is involved. Marin [**1**] uses this to give a stablized microbundle transversality, which is the "categorically correct" result because submanifolds have unique normal stabilized microbundles. These theorems apply except when the dimension of the ambient manifold is 4, or when one of the submanifolds has dimension 4 and the other has codimension 1 or 2, or when the expected dimension of the intersection is 4. This implies map transversality except when domain or preimage dimensions are 4.

A version of TOP map transversality with preimage dimension 4 was developed by Scharlemann [**1**].

The effect of the 4-dimensional development is to show TOP microbundle transversality holds in all dimensions and codimensions. Accordingly so do Marin's stabilized version, and map transversality. The previously excluded cases with ambient dimension different from 4 follow from the Kirby-Siebenmann proof and the submanifold smoothing theorem 8.7C (a sharpened version of the product structure theorem); see Quinn [**9**]. The ambient dimension 4 cases are given in 9.5.

CHAPTER 10

Classifications and Embeddings

Technically, the main result of the section is the π_1-negligible embedding theorem 10.6, which gives existence and uniqueness criteria for codimension zero embeddings of 4-manifolds with boundary. From this we deduce a criterion for a manifold to decompose as a connected sum. Perhaps the most basic result, the classification of closed simply-connected 4-manifolds, is obtained as a corollary of the sum theorem.

These results are presented in the reverse of the logical order because the more general ones are more complicated and technical. The chapter begins, then, with the classification theorem in 10.1. The invariants used in the classification are discussed in 10.2. Connected sums are described in 10.3, and the characterization theorem is deduced from this. π_1-negligible embeddings are discussed in 10.5, and using this the connected sum theorem is deduced in 10.6. A classification of manifolds with infinite cyclic fundamental group is given as an exercise in section 10.7. Finally the embedding theorem is proved; the existence part in 10.8, and the uniqueness using embeddings in 5-manifolds in 10.9. The technically inclined reader may prefer to see these in logical order, beginning with 10.5.

10.1 Classification of closed 1-connected 4-manifolds

For the classification theorem we recall that if M is a compact oriented manifold then intersection numbers define a bilinear form $\lambda \colon H_2M \otimes H_2M \to \mathbf{Z}$. This is symmetric, and if ∂M is empty it is nonsingular (the adjoint $H_2M \to (H_2M)^*$ is the Poincaré duality isomorphism). The Kirby-Siebenmann invariant $\mathrm{ks}M \in \mathbf{Z}/2$ is the obstruction to the existence of a smooth structure on $M \times \mathbf{R}$; see 8.3D and 10.2B below.

Theorem.

(1) **Existence:** *Suppose (H, λ) is a nonsingular symmetric form on a finitely generated free Z-module, $k \in Z/2$, and if λ is even then we assume $k \equiv (\text{signature}\lambda)/8$, mod 2. Then there is a closed oriented 1-connected manifold with form λ and Kirby-Siebenmann invariant k.*

(2) **Uniqueness:** *Suppose M, N are closed and 1-connected, $h\colon H_2$ $\to H_2 N$ is an isomorphism which preserves intersection forms, and $\mathrm{ks}M = \mathrm{ks}N$. Then there is a homeomorphism $f\colon M \to N$, unique up to isotopy, such that $f_* = h$.*

For example, S^4 and $S^2 \times S^2$ have even forms, so are uniquely determined by their forms. CP^2 has an odd form, so there is another manifold with this form but which is not stably smoothable.

Note that by setting $M = N$ in the uniqueness statement we see that taking a homeomorphism to its induced isometry of the form gives an isomorphism $\pi_0 \mathrm{TOP}(M) \to \mathrm{ISO}(H_2 M, \lambda)$. $\pi_0 \mathrm{TOP}(M)$ is the group of path components of the homeomorphism group of M, which is the same as isotopy classes of homeomorphisms. A homotopy equivalence also induces an isometry of the form, but the analogous statement for these is false; there are often many non-homotopic homotopy equivalences inducing each isometry (see Wall [**1**, §16], Quinn [**8**], and Cochran and Habegger [**1**]).

Some notation, suggested by the theorem, will be useful. If λ is a nonsingular form, denote by $\|\lambda\|$ the manifold with this form, and such that $\mathrm{ks}(\|\lambda\|) = 0$ if the form is not even. So for example $CP^2 = \|[1]\|$, and $S^2 \times S^2 = \left\|\left[\begin{smallmatrix} 0 & 1 \\ 1 & 0 \end{smallmatrix}\right]\right\|$. Part (1) of the theorem implies $\|\lambda\|$ exists, and (2) implies it is well-defined up to homeomorphism.

Suppose M is closed and 1-connected. Define $*M$ to be M if the form is even, and the manifold with the same form and opposite Kirby-Siebenmann invariant if the form is odd. Again the theorem implies this exists and is well-defined, and every closed 1-connected manifold is either $\|\lambda\|$ or $*\|\lambda\|$ for some appropriate λ. This operation will be defined (independently of the classification theorem) for a wider class of manifolds in 10.4.

10.2 The invariants

We briefly describe the invariants used in the theorem. For further information about forms over the integers see Milnor-Husemoller [**1**, Chapter II]. Hirzebruch and Neumann [**1**] describe relations between these invariants and the topology of manifolds. For information about the Kirby-Siebenmann invariant see Kirby-Siebenmann [**1**]. For a geometric point of view on the algebraic invariants of 4-manifolds see Kirby [**2**].

10.2A Symmetric forms over the integers. Important invariants of these are rank, signature (or "index"), type, and whether or not the form is definite.

The *rank* is the rank of the group on which the form is defined. The *type* is "even" if $\lambda(x, x)$ is even for all x, and is "odd" otherwise. The form is *definite* if $\lambda(x,x)$ is always nonnegative, or always nonpositive (and then is positive definite, or negative definite). Finally when the form is tensored with the real numbers it can be divided into positive and negative eigenspaces. The *signature* of the form is the dimension of the positive eigenspace minus the dimension of the negative eigenspace.

For example $\begin{bmatrix} 0 & 1 \\ 1 & 0 \end{bmatrix}$, the form of $S^2 \times S^2$, has rank 2, is even, is indefinite, and has signature 0. The form of CP^2 is [1], which has rank 1, is odd, positive definite, and has signature 1.

There are some simple relations among these invariants. The sum of the dimensions used to define the signature gives the rank. Therefore $|\text{signature}(\lambda)| \leq \text{rank}(\lambda)$, and $\text{signature}(\lambda) \equiv \text{rank}(\lambda) \bmod 2$. Similarly the form is definite if and only if the real form is definite, so if and only if $|\text{signature}(\lambda)| = \text{rank}(\lambda)$. A deeper fact is that if the form is even then the signature is divisible by 8.

Indefinite forms are determined up to isometry by the rank, signature, and type. If the form is odd, then it is isomorphic to $j[1] \oplus k[-1]$, where $j + k = \text{rank}$ and $j - k = \text{signature}$. If the form is even then it is isomorphic to $j(\pm E_8) \oplus k\begin{bmatrix} 0 & 1 \\ 1 & 0 \end{bmatrix}$, where $8j + 2k = \text{rank}$, $8j = |\text{signature}|$, and E_8 is the even definite form of rank 8 represented by the following matrix (blank entries are 0);

$$\begin{pmatrix} 2 & 1 & & & & & & \\ 1 & 2 & 1 & & & & & \\ & 1 & 2 & 1 & & & & \\ & & 1 & 2 & 1 & & & \\ & & & 1 & 2 & 1 & & 1 \\ & & & & 1 & 2 & 1 & \\ & & & & & 1 & 2 & \\ & & & & 1 & & & 2 \end{pmatrix}$$

An interesting family of examples is provided by the algebraic surfaces $Z_d \subset CP^3$ obtained from the zeros of a generic homogeneous complex polynomial of degree d. $H_2(Z_d)$ has rank $d^3 - 4d^2 + 6d - 2$, the form has signature $-\frac{1}{3}(d^3 - 4d)$, and is even if and only if d is even. According to the classification described above, this information is sufficient to identify these forms up to isomorphism. For example the Kummer surface $(d = 4)$ has rank 22, and even form with signature -16. Therefore the form is $2(-E_8) \oplus 3\begin{bmatrix} 0 & 1 \\ 1 & 0 \end{bmatrix}$.

Definite forms are much more complicated. Unlike the indefinite forms they have unique decompositions into sums of indecomposibles (Eichler's theorem, Milnor-Husemoller [1, Theorem II6.4]). Unfortunately there are over 10^{51} different indecomposable even definite forms of rank 40, and this number grows rapidly with increasing rank. We know nothing useful about them. None of this host of indecomposible definite forms occurs as the form of a smooth 4-manifold, however (see 8.4A).

If λ is the intersection form of a 4-manifold, then $\omega_2(x) \equiv \lambda(x,x)$, mod 2. Therefore evenness of the form is equivalent to vanishing of ω_2 on the integral homology $H_2(M; \mathbf{Z})$. If there is no 2-torsion in $H_1(M; \mathbf{Z})$ then this also implies ω_2 vanishes on mod 2 homology $H_2(M; \mathbf{Z}/2)$, which is equivalent to the existence of a spin structure. (For the significance of spin structures, see below.) We caution that without the H_1 condition the connection with spin structures can fail: there are non-spin manifolds with even forms on integral homology (Habegger [1]).

The form of a connected sum is the direct sum of the forms; $(H_2(M_1 \# M_2), \lambda) = (H_2(M_1), \lambda_1) \oplus (H_2(M_2), \lambda_2)$. The characterization theorem implies a converse; a direct sum decomposition of the form comes from a (topological) connected sum decomposition of the manifold. For example the Kummer surface decomposes as $2\| - E_8 \| \# 3S^2 \times S^2$. It does not decompose smoothly, however (see 8.4A).

If M is a compact manifold with boundary there is still an intersection form defined on $H_2(M; \mathbf{Z})$, but it usually is singular. All the invariants can be defined for singular forms, but only the signature seems to be useful. If there is an isomorphism of boundaries $\partial M \simeq \partial N$ then "Novikov additivity" asserts that signature$(M \cup_\partial N) = $ signature$(M) + $ signature(N), where the right side of the equation uses the signature of the singular forms. This generalizes the additivity for connected sums, where the boundary is S^3.

10.2B The Kirby-Siebenmann invariant. Suppose M is a compact topological 4-manifold. The boundary has a unique smooth structure, and there is an obstruction in $H^4(M, \partial M; \mathbf{Z}/2)$ to extension of this to a smooth structure on $M \times \mathbf{R}$, or $M \# kS^2 \times S^2$ (see 8.3D and 8.7). Each component of M has $H^4(-, \partial; \mathbf{Z}/2) \simeq \mathbf{Z}/2$. Define ks$(M)$ to be the sum over all components, of these invariants in $\mathbf{Z}/2$.

If M is connected then ks(M) is the stable smoothing obstruction. In general it is the number of components mod 2 which are not stably smoothable.

This number is an unoriented bordism invariant (i.e. if $M \cup N = \partial W$ then ks$(M) = $ ks(N)), and is additive in a very strong sense. If M is a union of manifolds $N_0 \cup N_1$ with $N_0 \cap N_1$ in the boundary of each N_i, then ks$(M) = $ ks$(N_0) + $ ks(N_1). This is a consequence of the uniqueness of smooth structures on 3-manifolds.

In general this invariant is somewhat distantly related to other characteristic classes, but there is a direct relation for spin manifolds. Let

SPTOP denote the universal cover of the identity component of the stable homeomorphism group TOP. (This is a 2-fold cover; π_0 and π_1 of TOP are the same as those of the orthogonal group O, and are both $\mathbf{Z}/2$). A *spin structure* is a lifting of the classifying map for the stable tangent bundle from B_{TOP} to B_{SPTOP}. A spin structure exists if $\omega_1 M = \omega_2 M = 0$ on $H_2(M; \mathbf{Z}/2)$. When one such structure exists others are classified by $H^1(M; \mathbf{Z}/2)$.

The basis for the relation is the theorem of Rochlin that a closed smooth spin 4-manifold has signature divisible by 16. Divisibility by 8 follows from the evenness of the form, so the force of the theorem is that there is another factor of 2. For closed topological spin manifolds the result is that signature$(M)/8 = \text{ks}(M)$ mod 2. Note this requires only the existence of a spin structure, not a particular choice.

Now we consider manifolds with boundary. Suppose N is a closed 3-manifold, with a spin structure denoted by τ. The 3-dimensional smooth spin bordism group is trivial, so (N, τ) is the boundary of a smooth spin 4-manifold (W, η). The *Rochlin invariant* $\text{roc}(N, \tau)$ is defined to be the signature of W, mod 16. This is well defined: Two such bounding manifolds glue together to give a closed smooth spin 4-manifold, whose signature is the difference of the signatures of the pieces. According to Rochlin's theorem for closed manifolds this difference is divisible by 16.

This invariant may depend on the spin structure. For example the 3-torus has $H^1(M; \mathbf{Z}/2)$ of order 8, so has 8 spin structures. 7 of these have invariant 0, and one (corresponding to the Lie group framing) has invariant 8 (see Kirby [2]). When there is a unique spin structure then of course the invariant is well defined.

This occurs when $H^1(N; \mathbf{Z}/2) = 0$, i.e. exactly when components of N are $\mathbf{Z}/2$ homology spheres.

Proposition. *Suppose (M, η) is a compact oriented 4-dimensional topological spin manifold. Then $8\text{ks}(M) \equiv \text{signature}(M) + \text{roc}(\partial M, \partial \eta)$, mod 16.*

This follows easily from the additivity of the signature and ks, and the closed case. When M is not spin there is a more complicated formula involving the Brown invariant of a linking form on a surface dual to ω_2, see Guillou and Marin [1], and Kirby [2].

10.3 Connected sum decompositions

The objective is to determine when a manifold W can be expressed as a connected sum $M \# W_1$, where M is a closed simply connected 4-manifold. The hypotheses are in terms of intersection and selfintersection forms on $\pi_2 W$.

If W is a connected sum $M \# W_1$, with M closed and 1-connected, then $\pi_2 W \simeq (\pi_2 M \otimes \mathbf{Z}\pi_1 W) \oplus \pi_2 W_1$. Intersection numbers on the M summand are given by: if $x \otimes \alpha$, $y \otimes \beta \in \pi_2 M \otimes \mathbf{Z}\pi_1 W$, then $\lambda(x \otimes \alpha, y \otimes \beta) = \lambda(x, y)\alpha\bar{\omega}(\beta)$. Similarly $\tilde{\mu}(x \otimes \alpha) = \tilde{\mu}(x)\alpha\bar{\omega}(\alpha)$.

Abstracting this, we say a $\mathbf{Z}\pi_1 W$ homomorphism $(\pi_2 M \otimes \mathbf{Z}\pi_1 W) \rightarrow \pi_2 W$ "preserves λ and $\tilde{\mu}$" if intersection numbers of images are given by these expressions. Since λ in M is nonsingular this implies the homomorphism is an injection onto a direct summand of $\pi_2 W$.

Theorem. *Suppose M is a closed 1-connected 4-manifold, and W has good fundamental group.*

    (1) *Let $(\pi_2 M) \otimes \mathbf{Z}\pi_1 W \rightarrow \pi_2 W$ be a $\pi_1 W$ monomorphism which preserves λ and $\tilde{\mu}$. If either $\omega_2 = 0$ on $\pi_2 W$ or ω_2 does not vanish on the subspace of $\pi_2 W$ perpendicular to the image, then there is a decomposition $W \simeq M \# W'$ inducing the given decomposition of π_2. If $\omega_2 \neq 0$ does vanish on the perpendicular then exactly one of W or $*W$ decomposes (see 10.4 below for $*W$).*

    (2) *Suppose $h_1 \colon W \simeq M \# W_1$ and $h_2 \colon W \simeq M \# W_2$ are two decompositions inducing the same decomposition of π_2. If $\omega_1 \colon \pi_1 W \rightarrow \mathbf{Z}/2$ is injective on elements of order 2, then the decompositions are pseudoisotopic. If the form of M is even the canonical homotopy equivalence $W_1 \rightarrow W_2$ is homotopic to a homeomorphism (regardless of ω_1).*

Two decompositions are isotopic if there is an isotopy from the identity to a homeomorphism $g \colon W \rightarrow W$ so that $h_2 g(h_1)^{-1}$ is the identity on M_0. Similarly they are pseudoisotopic if there is a pseudoisotopy to such a g; a homeomorphism $W \times I \simeq W \times I$ which is the identity on one end and g on the other. Pseudoisotopy implies isotopy if $\pi_1 W = 0$ (Quinn [**8**]).

We remark on the ω_2 hypothesis. If $J \subset \pi_2 W$ is a subgroup then a is "perpendicular" to J if $\lambda(a, b) = 0$ for all $b \in J$. Therefore the phrase "ω_2 vanishes on the subspace perpendicular to J" means that if $\lambda(a, b) = 0$ for all $b \in J$ then $\omega_2(a) = 0$. If the form λ is nonsingular on π_2, for example if W is closed, then this condition is equivalent to the existence of an element $b \in J$ dual to ω_2 in the sense that $\lambda(a, b) \equiv \omega_2(a)$, mod 2.

To complete the statement a definition of the $*$ operation used in the second part of (1) is needed. This was described in a special case in the discussion after the characterization theorem.

10.4 Definition of $*W$

Suppose W is a 4-manifold. If $\omega_2 \colon \pi_2 W \rightarrow \mathbf{Z}/2$ is trivial, define $*W =$

W. If ω_2 is nontrivial define $*W$ to be a manifold with a homeomorphism $(*W) \# CP^2 \simeq W \# (*CP^2)$ which preserves the decompositions of π_2, where $*CP^2$ is a manifold with form [1] and ks $= 1$.

Note that if $*W$ is defined and different from W, then it has the opposite Kirby-Siebenmann invariant. Also there is a canonical homotopy equivalence $*W \to W$ which is an isomorphism on the boundary.

The manifold $*CP^2$ is shown to exist and be unique in the beginning of the proof of 10.1, below. If the fundamental group of W is good then the first part of 10.3(1) implies that a manifold $*M$ exists: first note that $\pi_2 W$ is the subspace of $\pi_2(W \# *CP^2)$ perpendicular to $\pi_2(*CP^2) \otimes \mathbf{Z}\pi_1 W$. Therefore if $\omega_2 \neq 0$ on $\pi_2 W$, part (1) of the theorem implies the injection of $\pi_2(*CP^2) \otimes \mathbf{Z}\pi_1 W$ corresponds to a decomposition as a connected sum with CP^2. The complementary piece of the decomposition is—by definition—$*W$.

Similarly (2) shows that if $\pi_1 W$ is good and $\omega_1 \colon \pi_1 W \to \mathbf{Z}/2$ is injective on elements of order 2, then $*W$ is well-defined up to homeomorphism. This applies for example when $\pi_1 W$ has no 2-torsion, or is RP^4. We do not know if $*W$ is well-defined when the ω_1 condition fails, for example if W is $RP^3 \times S^1$ or $RP^2 \times RP^2$. ∎

Proof of 10.1: We deduce the Characterization Theorem from the sum theorem, and the fact that a homology 3-sphere bounds a contractible topological 4-manifold.

The first step is to realize the matrix E_8. This can be realized as the intersection matrix of a simply connected manifold with boundary, by plumbing together 8 copies of the D^2 bundle over S^2 whose core 2-sphere has selfintersection 2 (this is the tangent disk bundle of S^2). Specifically, form a linear chain of 7 copies by introducing single intersection points between their core spheres. Then introduce an intersection between the remaining one, and the third from one end in the chain.

It follows from the fact that the intersection matrix is nonsingular that the boundary of this manifold is a homology sphere (see Browder [1, V.2.6]; in fact it is the famous Poincaré homology sphere). According to 9.3C a homology sphere bounds a contractible manifold. The union of the plumbing manifold and the contractible one gives a closed 1-connected manifold which we denote by $\| E_8 \|$.

Next we construct $*CP^2$, with form [1] but ks $= 1$. According to 10.2B, since E_8 is even ks$\| E_8 \| = 1$, and by additivity ks$(\| E_8 \| \# (-CP^2)) = 1$. However according to the classification of indefinite forms, the form of this manifold is isomorphic to $8[1] \oplus [-1]$, which is the form of $8CP^2 \# (-CP^2)$. Restrict the isomorphism to get an injection of the form of $7CP^2 \# (-CP^2)$ to a direct summand. The perpendicular subspace is the homology of the remaining copy of CP^2, so the form is not even on it, and ω_2 does not vanish on it. The existence part of the decomposition theorem therefore applies to show there is a decomposition $(\|E_8\| \# (-CP^2)) \simeq 7CP^2 \# (-CP^2) \# N$, for some manifold N. This manifold has form [1], and additivity of ks shows ks$N = 1$. It therefore satisfies the conditions required of $*CP^2$.

Next we show there is a manifold realizing an arbitrary form λ. Since $\left[\begin{smallmatrix} 1 & 0 \\ 0 & -1 \end{smallmatrix}\right]$ is indefinite and odd, $\lambda \oplus \left[\begin{smallmatrix} 1 & 0 \\ 0 & -1 \end{smallmatrix}\right]$ is also. By the classification of forms it is isomorphic to $j[1] \oplus k[-1]$ for some j, k. This can be realized as the form of the manifold $jCP^2 \# k(-CP^2)$. The isomorphism gives an injection of $\left[\begin{smallmatrix} 1 & 0 \\ 0 & -1 \end{smallmatrix}\right]$ into this form. Since $\left[\begin{smallmatrix} 1 & 0 \\ 0 & -1 \end{smallmatrix}\right]$ is the form of $CP^2 \# (-CP^2)$, part (1) of the theorem gives a decomposition $jCP^2 \# k(-CP^2) \simeq CP^2 \# (-CP^2) \# N$, for some manifold N with form λ.

We now have all the manifolds required for the existence part of the characterization theorem; the paragraph above gives at least one with any given form, and the operation $*$ defined in 10.4 reverses the Kirby-Siebenmann invariant when the form is not even.

For the uniqueness part suppose M and N are closed and 1-connected, have the same Kirby-Siebenmann invariant, and $h\colon H_2M \to H_2N$ is an isomorphism which preserves forms. Regarding an isomorphism as an injection, the theorem asserts that there is a decomposition, either $N \simeq M \# P$, or $*N \simeq M \# P$, realizing this injection. However P must be a simply connected homology sphere, so is homeomorphic to S^4, by the Poincaré conjecture Corollary 7.1B. This means $N \simeq M$ or $*N \simeq M$. Since $*$ changes the Kirby-Siebenmann invariant if it changes the manifold, we conclude there is a homeomorphism $N \simeq N$ realizing the isomorphism of forms.

Finally we show that the homeomorphism is determined up to isotopy. Suppose $h_i\colon M \to N$ are homeomorphisms, $i = 1, 2$, which induce the same homomorphism on H_2. Regard N as $N \# S^4$, then according to the uniqueness part of the decomposition theorem there is a pseudoisotopy of the identity of M to a homeomorphism g so that $(h_1g)(h_2^{-1})$ is the identity on N_0. Since M is 1-connected, pseudoisotopy implies isotopy (Quinn, [8]). The complement of N_0 is a disk, and $h_1g(h_2^{-1})$ is the identity on the boundary. Therefore Alexander's isotopy ("squeeze toward the middle") on the disk gives an isotopy of $(h_1g)(h_2^{-1})$ to the identity. The Alexander isotopy and the isotopy of g therefore give an isotopy of h_1 to a map which when composed with (h_2^{-1}) gives the identity. This characterizes h_2, so h_1 is isotopic to h_2, as required for the theorem. ∎

10.5 π_1-negligible embeddings

An embedding $(V, \partial_0V) \to (W, \partial W)$ is π_1-negligible if $\pi_1(W - V) \to \pi_1(W)$ is an isomorphism. We also assume it is proper in the sense that there is a collar on ∂W which intersects V in a collar on $\partial_0 V$. The disk embedding theorems of chapter 5 provide π_1-negligible embeddings of unions of 2-handles $(D^2 \times D^2, S^1 \times D^2)$ extending a given embedding of $S^1 \times D^2$ in the boundary. Here this is generalized to embeddings of 4-manifold pairs $(V, \partial_0 V)$, extending given embeddings of $\partial_0 V$ in ∂W.

10.5A Embedding theorem. *Let* $(V; \partial_0 V, \partial_1 V)$ *be a compact 4-manifold triad so that* $\pi_1(V, \partial_0 V) = \{1\} = \pi_1(V, \partial_1 V)$ *(all basepoints), each component has nonempty intersection with* $\partial_1 V$, *and components disjoint from* $\partial_0 V$ *are 1-connected. Suppose* $h\colon V \to W$ *is a map which restricts to an embedding of* $\partial_0 V$ *in* ∂W.

(1) **Existence:** *Suppose* $\pi_1 W$ *is good,* $H_f^3(W_h, V \cup \partial W; \mathbf{Z}\pi_1 W) = 0$, *and* h *"preserves relative intersection and selfintersection numbers." If* ω_2 *is trivial on* $\pi_2 W$, *or does not vanish on the subspace of* $\pi_2 W$ *perpendicular to* $H_2(V, \partial_0 V; \mathbf{Z}\pi_1 W)$, *then* h *is homotopic rel* $\partial_0 V$ *to a* π_1-*negligible embedding. If* $\omega_2 \neq 0$ *does vanish on the perpendicular, then* h *is homotopic to such an embedding in exactly one of* W *or* $*W$.

(2) **Uniqueness:** *Suppose* $h_0, h_1\colon V \times I \to W$ *are* π_1-*negligible embeddings, homotopic rel* $\partial_0 V$. *Then there is an obstruction in* $H^2(V, \partial_0 V; (\mathbf{Z}/2)[T_+])$, *where* T_+ *is the set of elements in* $\ker \omega_2 \subset \pi_1 W$ *of order exactly two. If this vanishes then* h_0 *is* π_1-*negligibly concordant to* h_1.

A concordance of embeddings, as in (2), is an embedding $V \times I \to W \times I$ which restricts to the given embeddings on $V \times \{i\} \to W \times \{i\}$, $i = 0, 1$. The π_1-negligibility and duality imply that the complement of the interior of $F(V \times I)$ is an s-cobordism. If the fundamental group of W is good then the s-cobordism theorem implies this has a product structure, so the two embeddings are "pseudoisotopic." Finally if W is 1-connected the pseudoisotopy theorem of Quinn [**8**] implies the embeddings are isotopic. We caution that the concordance produced by the theorem may not be homotopic to the original homotopy.

The proof of the existence part of the theorem is given in 10.8, and the uniqueness in 10.9.

We now discuss the hypotheses. H_f^3 denotes the finite cochain cohomology, which on compact spaces is the same as ordinary cohomology. For manifolds it is Poincaré dual to ordinary homology even when the manifold is noncompact. W_h denotes the mapping cylinder of h. As in 10.3 above, the subspace of $\pi_2 W$ perpendicular to $H_2(V, \partial_0 V)$ is the set of elements $a \in \pi_2 W$ with $\lambda(a, b) = 0$ for all $b \in H_2(V, \partial_0 V; \mathbf{Z}\pi_1 W)$.

Next "preserves relative intersection numbers" will be defined. The term is supposed to suggest that relative classes in $(V, \partial_0 V)$ have the same algebraic intersection structure as the images in $(W, \partial_0 W)$. This cannot be defined the way it sounds because intersection numbers are not well defined for relative classes. Instead we form absolute classes from the differences $a - h_* a$ in $V \cup W$, and say that "relative intersections" are the same if the absolute intersections of these differences are trivial.

Let $d_h \colon H_i(V, \partial_0 V; \mathbf{Z}\pi) \to H_i(V \cup_{\partial_0 V} W; \mathbf{Z}\pi)$ be induced by the chain map which takes a relative chain x to $x - h_* x$. Define $\lambda_h(x, y)$ to be the intersection number $\lambda(d_h(x), d_h(y)) \in \mathbf{Z}\pi$, and similarly $\tilde{\mu}_h(x) = \tilde{\mu}(d_h(x))$. Then "$h$ preserves relative intersection numbers" if λ_h and μ_h vanish on $H_i(V, \partial_0 V; \mathbf{Z}\pi)$.

Sometimes the selfintersection part of this definition can be omitted. Recall that intersection numbers (in $4n$-manifolds) determine selfintersections except for coefficients of elements $g \in \pi_1 W$ with $g = g^{-1}$ and $\omega_1(g) = 1$. Therefore the selfintersection condition is unnecessary if ω_1 is trivial on 2-torsion in $\pi_1 W$. In general if the selfintersection hypothesis is dropped then the undetermined part can be organized to give an obstruction: denote the elements of order 2 and $\omega_1 = 1$ by T_-, then we get an obstruction in $H^2(V, \partial_0 V; \mathbf{Z}/2[T_-])$.

Analogously to this the $\mathbf{Z}/2[T_+]$ obstruction encountered in the uniqueness part of the theorem also comes from a selfintersection problem. The signs change because this problem occurs in a 6-manifold.

Next we show that many of the hypotheses are necessary as well as sufficient.

10.5B Proposition. *Suppose $h \colon V \to W$, as above, is homotopic rel $\partial_0 V$ to a π_1-negligible embedding, either in W or $*W$. Then $\pi_1(V, \partial_1 V) = \{1\}$ (for all basepoints), h "preserves relative intersection and selfintersection numbers," and $H_f^3(W_h, V \cup \partial W; \mathbf{Z}\pi_1 W) = 0$.*

Basically, the only condition in 10.5A that is not always necessary is the requirement that $\pi_1(V, \partial_0 V) = \{1\}$.

Proof: If V is embedded, the Seifert-Van Kampen theorem identifies the fundamental group of W as the free product of $\pi_1 V$ and π_1 of the complement, amalgamated over $\pi_1 \partial_1 V$. If the complement has the same fundamental group as W then $\partial_1 V$ must be connected and $\pi_1 \partial_1 V \to \pi_1 V$ must be onto. The long exact sequence of the pair shows these to be equivalent to the vanishing of $\pi_1(V, \partial_1 V)$.

An embedding does not change intersections, so preserves intersection numbers. Finally if h is an embedding then by excision $H_f^3(W_h, V \cup \partial W; \mathbf{Z}\pi_1 W) = H_f^3(W - \operatorname{int} h(V), \partial(W - \operatorname{int} h(V)))$. This is dual to $H_1(W - \operatorname{int} h(V); \mathbf{Z}\pi_1 W)$, which is H_1 of the covering space associated to the homomorphism to $\pi_1 W$. The π_1-negligible condition implies this covering space is 1-connected, so H_1 vanishes. ∎

10.6 Proof of 10.3

We show that the connected sum decomposition theorem follows from the π_1-negligible embedding theorem.

Let M be a closed simply-connected manifold, and denote—as above— the complement of a flat open 4-ball by M_0. Consider this as a triad with $\partial_0 M_0$ empty, then a decompostion of W as a connected sum with M corresponds exactly to an embedding of this triad in W.

M_0 is homotopy equivalent to a wedge of 2-spheres, so $\pi_2 M \otimes \mathbf{Z}\pi_1 W$ is a free $\mathbf{Z}\pi_1 W$-module. Given a $\mathbf{Z}\pi_1 W$ homomorphism $(\pi_2 M \otimes \mathbf{Z}\pi_1 W) \to \pi_2 W$ there is a map (unique up to homotopy) $M_0 \to W$ inducing it. This map preserves intersection and selfintersection numbers in the sense of 10.5 if and only if the homomorphism preserves λ and μ in the sense of 10.3. Finally the ω_2 on perpendicular subspaces is the same in the two statements. Thus 10.5(1) implies 10.3(1).

For the uniqueness part, assume that two connected sum decompositions of W are given. Since the homomorphism on π_2 determines the map $M_0 \to W$ up to homotopy, if the decompositions induce the same decomposition of $\pi_2 W$ the embeddings are homotopic. Therefore 10.5(2) provides an obstruction in $\mathbf{Z}/2[T_+]$ whose vanishing implies pseudoisotopy of the embeddings. But T_+ is the set of elements in the kernel of $\omega_1 \colon \pi_1 W \to \mathbf{Z}/2$ of exponent exactly 2, so the injectivity assumption in 10.3(2) is equivalent to T_+ being empty. Thus 10.5(2) implies the pseudoisotopy part of 10.3(2).

To complete the proof of 10.3 we need uniqueness of the complement when the form of M is even. This is proved independently of 10.5, and in fact will be used in its proof.

We suppose $W \simeq M \# W_1 \simeq M \# W_2$. Define $\hat{W}$ by connected sum of with a copy of $-M$, so $\hat{W} \simeq \hat{M} \# W_1 \simeq \hat{M} \# W_2$, with $\hat{M} = M \# (-M)$. The form of $\hat{M}$ is even, indefinite, and has signature 0, so it is isomorphic to $k\begin{bmatrix} 0 & 1 \\ 1 & 0 \end{bmatrix}$ for some k. According to the characterization theorem the manifold is supposed to be $kS^2 \times S^2$. We give a direct proof of this by finding the core spheres using the disk embedding theorem.

Choose a basis for $\pi_2\hat{M}$, $\{a_i, b_i\}$ for $i = 1, \dots, n$ so that all intersections and selfintersections are trivial except $\lambda(a_i, b_i) = 1$. Represent the $\{a_i\}$ by framed immersions $A_i \colon S^2 \to \hat{M}$, and choose a 4-ball D which intersects the A_i in mutually disjoint embedded disks. Delete the interior of D, then the result is a collection of framed immersed disks with algebraically transverse spheres (the b_i). $\hat{M}$ is simply connected, so the corollary to 5.1 applies to give a regular homotopy rel boundary of $\bigcup_i A_i$ to an embedding. Replacing D gives disjoint framed embedded spheres representing $\{a_i\}$.

Next find disjoint framed embedded spheres $\{B_i\}$ representing $\{b_i\}$, and so that the only intersections with the A_* is a single point in each $A_i \cap B_i$. This is done either by deleting a neighborhood of $\bigcup_i A_i$ and applying corollary 5.1 again, or much more directly by using the spheres A_i as in the exercise in 1.9.

Now let $h_j \colon \hat{W} \simeq \hat{M} \# W_j$ denote the given homeomorphism for $j = 1, 2$, so that $h_1 h_2^{-1}$ preserves the decompositions of the forms and is the identity on the part coming from $\hat{M}$. Construct a 5-manifold Z by starting with $\hat{W} \times I$, adding 3-handles on the spheres $h_1^{-1} A_i \subset \hat{W} \times \{0\}$, and 3-handles on the spheres $h_2^{-1} B_i \subset \hat{W} \times \{1\}$. The boundary of Z is the union of three pieces; W_1 connected sum the manifold obtained by surgery on $\bigcup_i A_i \subset \hat{M}$, W_2 connected sum the manifold obtained by surgery on $\bigcup_i B_i \subset \hat{M}$, and $(\partial W) \times I$. Since $\hat{M}$ is reduced to a homotopy sphere (thus a sphere) by surgery on either the A_i or the B_i, the first two pieces are W_1 and W_2.

The basic idea is that Z is an s-cobordism rel boundary from W_1 to W_2, so since the fundamental group is good the s-cobordism theorem implies the ends are homeomorphic. We have not assumed W is compact, so we use the "enlargement" version 7.1C of the s-cobordism theorem rather than 7.1A. There is a handlebody structure for (Z, W_1) with 2-handles the duals of the handles added to the $h_1^{-1} A_i$, and with 3-handles the handles added to the $h_2^{-1} B_i$. $\hat{W}$ is the level manifold between the 2- and 3-handles. Since $h_1 h_2^{-1}$ is the identity on the part of the forms coming from $\hat{M}$, the matrix of intersection between the $h_1^{-1} A_*$ and the $h_2^{-1} B_*$ is the identity.

Choose nullhomotopies for Whitney circles for all excess intersections and intersections, and let X be a compact submanifold of $\hat{W}$ containing all the spheres and nullhomotopies. Let Y be the submanifold of Z obtained by attaching the handles, then $(Y, Y \cap W_1)$ is an s-cobordism and Z has a product structure in the complement of Y. Since $\pi_1 Z = \pi_1 W$ is good, version 7.1C of the h-cobordism theorem implies that Z has a product structure. In particular there is a homeomorphism $W_1 \simeq W_2$, as required. ∎

10.7 Manifolds with infinite cyclic fundamental group

Analogs of the characterization and connected sum theorems are stated without formal proof, basically as an extended exercise. We denote the infinite cyclic group by J, to distinguish it from the coefficient ring $\mathbf{Z}$.

Suppose M is a closed manifold with $\pi_1 M \simeq J$. Then $H_2(M; \mathbf{Z}[J])$ is a free $\mathbf{Z}[J]$ module; the fact that it is stably free is a byproduct of the proofs of the results below, and for this ring stably free implies free.

When M is orientable the intersection form is a nonsingular hermitian form on this free module. Generally the form is defined using an orientation at the basepoint, and is $\bar{\omega}_1$-hermitian.

10.7A Classification Theorem.

(1) **Existence:** *Suppose (H, λ) is a nonsingular hermitian form on a finitely generated free $Z[J]$-module, $k \in Z/2$, and if λ is even then we assume $k \equiv (\text{signature}\lambda)/8$, mod 2. Then there is a oriented closed manifold with $\pi_1 = J$, intersection form λ and Kirby-Siebenmann invariant k.*

(2) **Uniqueness:** *Suppose M, N are closed and oriented with $\pi_1 = J$, $h \colon H_2(M; \mathbf{Z}[J]) \to H_2(N; \mathbf{Z}[J])$ is a $\mathbf{Z}[J]$ isomorphism which preserves intersection forms, and $\mathrm{ks}M = \mathrm{ks}N$. Then there is a homeomorphism $f \colon M \to N$, unique up to pseudoisotopy, which induces the given identification of fundamental groups, preserves orientation and such that $f_* = h$.*

This is the analog of 10.1, except the uniqueness is only up to pseudoisotopy. We caution that the situation with other fundamental groups is much more complicated (see 10.10).

To state the analog of the sum theorem we need an S^1 analog of connected sums. A map $S^1 \to M^4$ is homotopic to an embedding with neighborhood a D^3 bundle over S^1. This bundle is either $S^1 \times D^3$ or a Mobius band times D^2, depending on the value of ω_1 on the homology class. Now suppose there are embeddings of S^1 in both M and W, on which the respective ω_1 take the same value. Define $M \#_{S^1} W$ by deleting the interiors of disk bundle neighborhoods and identifying the boundaries. More precisely if local orientations are given then this can be done so the result has a local orientation compatible with those of the pieces, and when normalized in this way the operation is well-defined.

Continuing with the same notation, suppose M has fundamental group J and $S^1 \subset M$ represents the generator. Suppose also the inclusion $S^1 \subset N$ is injective on π_1. Then $\pi_1(M \#_{S^1} N) \simeq \pi_1 N$. Denoting this group by π, it is also true that $H_2(M \#_{S^1} N; \mathbf{Z}[\pi]) = (H_2(M; \mathbf{Z}[J]) \otimes_{\mathbf{Z}J} \mathbf{Z}[\pi]) \oplus H_2(N; \mathbf{Z}[\pi])$. Similarly the intersection form of the sum is the form on M tensored up to $\mathbf{Z}[\pi]$, plus the form on N.

10.7B Theorem.
Suppose M is a closed locally oriented 4-manifold with fundamental group J, W has good fundamental group π, $J \to \pi$ is injective, and ω_1 of the two manifolds agree on J.

(1) *Suppose $H_2(M; \mathbf{Z}[J]) \otimes_{\mathbf{Z}J} \mathbf{Z}[\pi] \to H_2 W$ is a $\mathbf{Z}\pi$ homomorphism which preserves λ and $\tilde{\mu}$, and either $\omega_2 = 0$ on $\pi_2 W$, or ω_2 does not vanish on the subspace perpendicular to the image. Then*

there is a decomposition $W \simeq M \#_{S^1} W'$ inducing the given homomorphism to $\pi_2 W$. If $\omega_2 \neq 0$ does vanish on the perpendicular, then exactly one of W or $*W$ decomposes.

(2) Suppose $h_1: W \simeq M \#_{S^1} W_1$ and $h_2: W \simeq M \#_{S^1} W_2$ are two decompositions inducing the same decomposition of π_2. If $\omega_1: \pi_1 W \longrightarrow \mathbf{Z}/2$ is injective on elements of order 2, then the decompositions are pseudoisotopic.

Note that, unlike the previous theorem, the manifolds are not assumed orientable.

This result is proved from the π_1-negligible embedding theorem. Denote by M_S the complement of the open disk bundle of the embedding $S^1 \subset M$, then the data of (1) defines a map $M_S \to W$. Let $S^1 \subset M_S$ be another generating circle, with disk bundle neighborhood E. Show that the map $M_S \to W$ is homotopic to one which is a homeomorphism on E and takes $M_S - E$ into the complement of the image of E. Then 10.5 applies to the map $(M_S - \mathrm{int} E, \partial E) \longrightarrow (W - \mathrm{im} E, \mathrm{im} \partial E)$.

It may be helpful to note that after addition of a closed 1-connected manifold with even form, $(M_S - \mathrm{int} E, \partial E)$ has a handlebody structure with only 2-handles (see the next section).

The characterization theorem is proved, as in the 1-connected case, from the sum theorem and a few additional facts. The first fact is a stable classification of nonsingular hermitian forms over $\mathbf{Z}[J]$. Given (H, λ) there is an $n \times n$ hermitian $\mathbf{Z}[J]$-matrix A (possibly singular) such that $(H, \lambda) \oplus (\mathbf{Z}[J]^{2n}, \left[\begin{smallmatrix} 0 & 1 \\ 1 & A \end{smallmatrix}\right])$ is isomorphic to a sum of copies of $[1]$ and $[-1]$. This is the statement that the group of stable equivalence classes of such forms is generated by the form $[1]$, see Ranicki [1, §10].

To remove summands $\left[\begin{smallmatrix} 0 & 1 \\ 1 & A \end{smallmatrix}\right]$ as in the proof of the 1-connected version, we need to construct manifolds with these forms. Begin with $E = S^1 \times D^3$. Choose n pairs of embedded disks B_i, C_i in ∂E which are mutually disjoint except for a clasp of intersections in $B_i \cap C_i$. Adding 2-handles to E on the standard framings of ∂B_i, ∂C_i gives $E \# n(S^2 \times S^2)$, with form $\left[\begin{smallmatrix} 0 & 1 \\ 1 & 0 \end{smallmatrix}\right]$. Now modify the C_i: push a piece of C_i around a loop g and introduce a clasp with C_j.

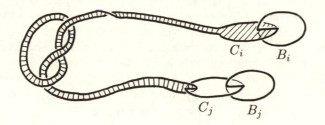

This changes the intersection matrix by the symmetrization of the matrix with $\pm g$ in the i, j place. Any symmetric matrix A over $\mathbf{Z}[J]$ can be realized by this operation and changing the framing on the ∂C_i. Attaching handles to these new framed circles gives a manifold with boundary, and form $\begin{bmatrix} 0 & 1 \\ 1 & A \end{bmatrix}$.

The nonsingularity of the form implies the boundary of this manifold has the $\mathbf{Z}[J]$ homology of ∂E. However (according to surgery, in 11.6) any such manifold bounds a topological manifold with the homotopy type of S^1. Attach such a bounding manifold to the construction to obtain a closed manifold with the desired form.

The uniqueness statements require an additional fact from surgery. Let E be a D^j bundle over S^1. If $N \to E$ is a homotopy equivalence which is a homeomorphism on the boundary, then it is homotopic rel boundary to a homeomorphism (see 11.5). This is used in the cases $j = 3$ and $j = 4$.

10.8 Proof of 10.5(1)

In outline, the theorem is first reduced to the case where $(V, \partial_0 V)$ is a handlebody with only 2-handles. The 2-handles are then embedded using the embedding theorem for disks. When $\omega_2 \neq 0$ on π_2 but vanishes on the subspace perpendicular to $H_2(V, \partial_0 V; \mathbf{Z}\pi_1 W)$ then complications arise because we cannot always find framed immersed dual classes for the 2-handles.

Much of the chapter to this point has consisted of deductions from the result now being proved. However the uniqueness of cancellation for manifolds with even forms, and the existence of the manifold $\| E_8 \|$ were established independently, in 10.6 and the proof of 10.1, respectively. We will use these facts here.

Suppose M is a closed 1-connected manifold with even form, and suppose the map obtained by connected sum with the identity $M \# V \to M \# W$ is homotopic rel $\partial_0 V$ to a π_1-null embedding. This gives a second embedding of M_0 in $M \# W$, and therefore a decomposition $M \# W \simeq M \# W'$, with the two copies of M representing the same summand of the form. The uniqueness of cancellation shows the canonical homotopy equivalence $W \simeq W'$ is homotopic rel ∂W to a homeomorphism. But W' has V embedded π_1-negligibly in it. Therefore we get an embedding of V in W.

This shows it is sufficient to embed $M \# V$ in $M \# W$, where M is any closed manifold with even form and 1-connected components.

As a first consequence we note we can assume V is smoothable. Add a copy of the E_8 manifold to each component with nontrivial Kirby-Siebenmann invariant to change it to be zero. According to 8.7(1), a manifold with vanishing Kirby-Siebenmann invariant can be smoothed after connected sum with some number of copies of $S^2 \times S^2$. Therefore $V \# j \| E_8 \| \# k(S^2 \times S^2)$ is smoothable, and to embed V it is sufficient to embed this.

The second consequence of cancellation is that we can assume $(V, \partial_0 V)$ has a handlebody structure with only 2-handles. We first arrange that $\partial_0 V$ intersects each component nontrivially; in each component where this fails make h a homeomorphism on a ball, delete the interior from both V and W, and add the boundary to $\partial_0 V$. Now since V is smoothable it has some handlebody structure. Each component intersects $\partial_0 V$ and $\partial_1 V$ nontrivially, so there is a handlebody structure with no 0- or 4-handles. 1- and 3-handles will be eliminated by showing they can be changed into 2-handles by connected sums with $S^2 \times S^2$.

Consider a 1-handle, attached to $\partial_0 V$. The connectivity assumptions imply that $\pi_1(V, \partial_0 V) = 0$, so the core of this 1-handle is homotopic relative to the ends into $\partial_0 V$.

Approximate the homotopy by a framed immersed 2-disk which is standard near the core of the handle, and push selfintersections off by finger moves (see 1.5). This gives a framed embedded disk whose boundary is the union of the handle core and an arc on $\partial_0 V$. Let B denote a 3-ball neighborhood of the arc in $\partial_0 V$, then a collar on B union with the 1-handle is isomorphic to $S^1 \times D^3$.

The connected sum operation $\# S^2 \times S^2$ removes a 4-ball and replaces it with $S^2 \times S^2 - D_-^2 \times D_-^2$, where $S^2 = D_-^2 \cup D_+^2$. Identify $D_-^2 \times D_+^2$ in this with a 1-handle in the original 4-ball. The complement of this 1-handle is isomorphic to $S^1 \times D^3$, so the operation becomes: remove a copy of $S^1 \times D^3$ (whose core bounds an embedded D^2), and replace it with $D^2 \times S^2$. (This is a "surgery" on the S^1.) Therefore by such a connected sum we can replace the 1-handle on $B \times I$ with a copy of $D^2 \times S^2$. This can be regarded as a 2-handle added on $B \times I$, so the original 1-handle has been replaced by a 2-handle.

A 3-handle can be considered dually as a 1-handle attached to $\partial_1 V$, so the same argument can be used to convert all 3-handles into 2-handles.

We now assume $(V, \partial_0 V)$ has only 2-handles. Denote the core 2-disks of these handles by $\{D_i\}$. The first step toward applying the disk embedding theorems is to find an appropriate immersion homotopic to h.

According to the immersion lemma h is homotopic to an immersion which differs by rotations from the given one on the attaching region of the handles. Let χ_i denote the rotation on $\partial_0 D_i$. $A_i = D_i \cup h(D_i)$ defines immersed 2-spheres with possibly nontrivial normal bundles. The 1 coefficient (in $\mathbf{Z}\pi_1 W$) of the selfintersection forms of these immersions satisfy $2\mu_1(A_i) + \chi_i = \lambda_1(A_i, A_i)$.

The hypothesis that h "preserves relative intersection numbers" implies $\lambda_1(A_i, A_i) = 0$, so χ_i is even. According to 1.3 it can be changed by any even number by twisting inside $h(D_i)$, so we can arrange for it to be zero. This gives an immersion of V extending the embedding of $\partial_0 V$, and the intersection and selfintersection forms of the core disks are trivial. Note that according to Proposition 1.7 this immersion is well-defined up to regular homotopy; different regular homotopy classes rel boundary of immersions of disks can be distinguished by selfintersection numbers.

Next we locate homologically transverse elements for the $h(D_i)$, in $\pi_2 W$.

Since the cohomology group $H^3_f(W_h, \partial W \cup V)$ ($\mathbf{Z}\pi_1 W$ coefficients) is assumed trivial, the long exact sequence of the triple $(W_h, \partial W \cup V, \partial W)$ shows that $H^2_f(W, \partial W) \to H^2(V, \partial_0 V)$ is onto. The first group is Poincaré dual to $H_2(W; \mathbf{Z}\pi_1 W) \simeq \pi_2 W$. Since $(V, \partial_0 V)$ has only 2-handles the cohomology group is isomorphic to the dual of the homology; $(H_2(V, \partial_0 V))^*$. The resulting homomorphism $H_2(W; \mathbf{Z}\pi_1 W) \to (H_2(V, \partial_0 V))^*$ is the adjoint of the intersection form $\lambda \colon H_2(W; \mathbf{Z}\pi_1 W) \times H_2(V, \partial_0 V) \to \mathbf{Z}\pi_1 W$. Since the homomorphism is onto we conclude that there are elements $\alpha_i \in \pi_2 W$ such that $\lambda(D_i, \alpha_j) = 0$ if $i \neq j$, and $= 1$ if $i = j$.

To apply the embedding theorem we need such classes α_i represented by framed immersions. According to 1.3 this will be the case if $\omega_2(\alpha_i) = 0$. If $\omega_2 = 0$ on $\pi_2 W$ this holds automatically. If there is an element $\gamma \in \pi_2 W$ so that $\lambda(D_i, \gamma) = 0$ for all i and $\omega_2(\gamma) \neq 0$, then we can add γ to the α_i with nontrivial ω_2, to obtain new α_i on which ω_2 vanishes. Therefore framed immersions can be obtained unless ω_2 is nontrivial on $\pi_2 W$, but vanishes on the subspace perpendicular to $H_2(V, \partial_0 V)$.

When there are framed immersed α_i, Corollary 5.1B implies that h is regularly homotopic rel $\partial_0 V$ to a π_1-negligible embedding. This proves theorem 10.5(1) unless $\omega_2 \neq 0$ on $\pi_2 W$, but vanishes on the perpendicular to $H_2(V, \partial_0 V)$.

To analyse the situation when the α_i cannot be chosen to be framed, we recall the proof of the appropriate embedding theorem (the proof of Corollary 5.1B from Theorem 5.1A). First immersed Whitney moves are used to change h by regular homotopy so that the α_i are represented by framed immersions which are transverse spheres for the $h(D_i)$. Then immersed Whitney disks are chosen for all intersections among the $h(D_i)$. These are made disjoint from the $h(D_i)$ by adding copies of the α_i. Framed immersed transverse spheres for the Whitney disks are constructed by contraction from the linking tori and caps, after the caps are made disjoint from $h(D_*)$ using the α_i. Theorem 5.1 then applies to the Whitney disks; they can be replaced by embeddings with the same framed boundaries. Whitney isotopies using these embedded disks give a regular homotopy of h to the desired embedding.

The only step which requires the framing is the use of sums with the α_i to make the Whitney disks disjoint from $h(D_i)$. For example, in the last step (constructing transverse spheres for the Whitney disks) each α_i enters algebraically zero times, because it is added to caps in a capped surface which is then contracted. This implies twists in the normal bundle cancel out. Since the problem comes from these Whitney disk–$h(D_i)$ intersections, we use them to define an invariant.

10.8A Definition. In the situation above suppose B is an immersed Whitney disk for intersections among the $h(D_*)$, with boundary arcs on $h(D_i)$ and $h(D_j)$. Define $\mathrm{km}(B) = \omega_2(\alpha_i)\omega_2(\alpha_j)\sum_k |B \cap h(D_k)|\omega_2(\alpha_k)$, in $\mathbf{Z}/2$. Here $|B \cap h(D_i)|$ denotes the number of intersection points, in general position, mod 2. If $\{B_j\}$ is a complete set of Whitney disks (with disjoint boundaries) for all the intersections, define $\mathrm{km}(h) = \sum_j \mathrm{km}(B_j)$. Finally if h_0 is a map as originally specified in the theorem, and h is a map of a 2-handlebody obtained from it as above, then define $\mathrm{km}(h_0) = \mathrm{km}(h)$.

This invariant extends an embedding obstruction for 2-spheres discovered by Kervaire and Milnor [1], and the notation is intended to reflect this. The following lemma completes the proof of part (1) of the theorem.

10.8B Lemma. *Suppose h satisfies the hypotheses of 10.5(1), and determines ω_2 on $\pi_2 W$. Then $\mathrm{km}(h)$ is well defined, and h is homotopic rel $\partial_0 V$ to an embedding in W, or $*W$, if and only if $\mathrm{km}(h) = 0$, or $\mathrm{km}(h) = 1$ respectively.*

Proof: We claim it is sufficient to show the invariant is well-defined, and that if it vanishes then h is homotopic to an embedding. Since an embedding trivially has vanishing invariant these assertions directly imply that h is homotopic to an embedding if and only if the invariant vanishes. We show next the assertions also imply that if $km(h) = 1$ then h is homotopic to an embedding in $*W$. Together these statements imply the lemma.

Let $f \colon CP^2 \to *CP^2$ denote the canonical homotopy equivalence, then $f_0 \colon CP_0^2 \to *CP^2$ satisfies the hypotheses of the theorem, but cannot be homotopic to an embedding. Therefore $km(f_0) = 1$. Consider $h \cup f_0 \colon V \cup CP_0^2 \to W \# (*CP^2)$. If $km(h) = 1$ then $km(h \cup f_0) = 0$, so is homotopic to an embedding. This gives an embedding of V in the complement of $CP_0^2 \subset W \# (*CP^2)$, which is by definition $(*W)_0$. Therefore h is homotopic to an embedding in $*W$.

We now show that if V has a handlebody structure with only 2-handles such that $km(h)$ (defined with respect to this structure) vanishes, then h is homotopic to an embedding.

The first step is an operation which changes intersections of Whitney disks. Suppose D is an immersed disk, and B, C are Whitney disks with disjoint boundaries, for intersections of D with a surface D'. Push a piece of the boundary arc of B through one of the C intersection points, to introduce a new BD' intersection and a point of intersection of the Whitney arcs.

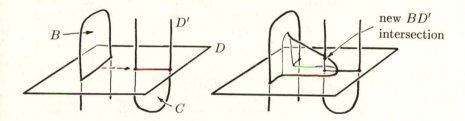

Now push the C boundary arc along the B arc through one of the B intersection points.

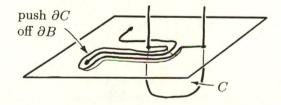

This makes the boundary arcs disjoint once again, and introduces a CD' intersection. If km is defined using intersections with D' then this procedure changes both $km(B)$ and $km(C)$.

Now suppose h is an immersion of a 2-handlebody, with Whitney disks $\{B_j\}$ and $km(h) = 0$. We change this so that each Whitney disk has $km = 0$. If there is a Whitney disk with $km \neq 0$ then there must be two since the total is 0 mod 2. Suppose they are B_1 and B_2, with a boundary arc on $h(D_1)$, $h(D_2)$ respectively. Use a finger move to introduce a new pair of intersections between $h(D_1)$ and $h(D_2)$, with embedded Whitney disk C with boundary arcs on both $h(D_1)$ and $h(D_2)$. Apply the procedure above to the pair C, B_1, and to C, B_2. A single new $B \cap h(D)$ intersection is introduced into each of B_1 and B_2, so they now have $km = 0$. Two intersection points are introduced into C, so it also has $km = 0$. This reduces the number of $km \neq 0$ Whitney disks by 2, so by induction they can all be converted.

Thus we may assume all Whitney disks have $km = 0$. Use connected sums with the (possibly unframed) transverse spheres to make them disjoint from the $h(D_*)$. If a disk B_j is changed by an even number of sums with spheres with $\omega_2 \neq 0$ then the normal bundle of the result differs on the boundary by an even number of twists from the Whitney framing. Interior twists as in 1.3 can be used to correct this. If there are an odd number of such spheres then by definition of $km(B_j)$ one of the boundary arcs must be on a disk with framed transverse sphere, say $h(D_i)$. A boundary twist about this arc corrects the framing on the Whitney disk, but introduces a new intersection with $h(D_i)$. Since $h(D_i)$ has framed transverse sphere this intersection can be removed without disturbing the framing.

Therefore we can arrange that there are Whitney disks with interiors disjoint from the $h(D_*)$. As noted above the rest of the proof of the embedding theorem does not require framings on the transverse spherse of the $h(D_*)$, so (if the fundamental group is good) there are embedded Whitney disks, and h is homotopic to an embedding.

The lemma will be completed by showing that $km(h)$ is well-defined. Note that if it is well-defined with respect to a fixed handlebody structure then its vanishing is equivalent to embedibility of the manifold. But embedibility is independent of the handlebody structure, and independent of the connected sums used to get such a structure. Therefore it is a well-defined invariant of the original map. This reduces the problem to showing it is well defined with respect to a fixed handlebody structure.

Fix a handlebody structure for $(V, \partial_0 V)$ with only 2-handles. Recall that the immersion homotopic to h is well-defined up to regular homotopy by the intersection number conditions. A regular homotopy is a sequence of Whitney moves, finger moves, and isotopies. Since a Whitney move is inverse to a finger move, it is sufficient to see that km is invariant under finger moves.

A finger move introduces a single embedded Whitney disk disjoint from everything else, so adding it to a collection of Whitney disks gives a new collection with the same invariant. This reduces the proof of the lemma to showing the invariant is independent of the choice of Whitney disks, for a fixed immersion in general position.

The first step is to generalize the definition. Define "weak" Whitney disks to be disks satisfying all the conditions on Whitney disks, except the boundary circles are allowed to be immersed arcs in $h(D_*)$. (Usually they are required to be disjointly embedded.) Now suppose $\{B_k\}$ is a complete collection of weak Whitney disks for the intersections of $h(D_*)$. Define $\mathrm{km}(B_*)$ to be the sum over all intersections $\mathrm{int}B_* \cap h(D_*)$ and selfintersections of ∂B_* of products $\omega_2(\alpha_i)\omega_2(\alpha_j)\omega_2(\alpha_k)$ in $\mathbf{Z}/2$. For a BD intersection, i is the index of the D_*, and j, k are the indices of the $h(D_*)$ containing the boundary of B. For a ∂B intersection between boundaries of distince B, i is the index of the D_* on which the intersection takes place, and j, k are the indices of D_* containing the other boundary arcs of the Bs. For a ∂B selfintersection, i is the index of the D_* on which the intersection takes place, and $j = k$ is the index of the D_* containing the other boundary arc of B.

If the Whitney disks are standard, so there are no boundary intersections, this reduces to the previous definition.

The new definition of km is invariant under regular homotopy of the Whitney disks. During a regular homotopy intersection points appear and disappear in pairs, leaving the sum in $\mathbf{Z}/2$ unchanged, except when one boundary arc of a Whitney disk passes over an endpoint of another boundary arc. This changes the number of boundary intersections by one. Since endpoints are $h(D_*)$ intersection points, it also changes the interior intersections by one, as in the pictures above. The $\mathbf{Z}/2$ elements assigned to the two types of intersections are the same, so the sum is invariant in this case also.

Next we show the invariant depends only on the Whitney circles, not on the choice of disks. Suppose B and B' are Whitney disks with the same boundary circle, on disks $h(D_i) \cup h(D_j)$. Both B and B' define framings of the normal bundle of the boundary circle. Using the uniqueness of the normal bundle of the circle, we may assume after isotopy that B' intersects the B framing in a subbundle, which may be twisted. Suppose there are r twists along the arc on $h(D_i)$ and s twists on $h(D_j)$.

Applying the boundary twisting operation of 1.3 to B' straightens out

B', but introduces rotations in its normal bundle. Once B and B' are aligned near the boundary they can be glued together a short distance out from the boundary curve to give an immersed sphere, C. Because of the rotations introduced by the twisting operation $\omega_2(C) = r + s$ mod 2.

Note that $\omega_2(C)$ is also given by $\sum_k |C \cap h(D_k)|\omega_2(\alpha_k)$. This is because C can be expressed (in $\pi_2 W$) as a linear combination of the α_*, plus an element perpendicular to $h(D_*)$. Since ω_2 vanishes on the perpendicular subspace, this last has $\omega_2 = 0$. The number of copies of α_k in the sum is $|C \cap h(D_k)|$, so the formula follows from additivity of ω_2.

This expression for $\omega_2(C)$ can be related to the km contributions of B and B'. C has the same intersections with the $h(D_*)$ as $\text{int}B \cup \text{int}B'$, plus some introduced by the twisting operation; r with $h(D_i)$ and s with $h(D_j)$. Since we have assumed $\omega_2(\alpha_i)$ and $\omega_2(\alpha_j)$ are nontrivial, the intersection expression gives $\omega_2(C) = \sum_k |(\text{int}B) \cap h(D_k)|\omega_2(\alpha_k) + \sum_k |(\text{int}B') \cap h(D_k)|\omega_2(\alpha_k) + r + s$.

Subtracting $\omega_2(C) = r + s$ and multiplying by $\omega_2(\alpha_i)\omega_2(\alpha_j)$ shows that the interior contributions of B and B' to km are equal. Therefore $\text{km}(B_*)$ depends only on ∂B_*.

Now we show it does not depend on ∂B either. Note that two collections of immersed arcs in a 2-disk, with the same endpoints, are regularly homotopic rel endpoints. Regular homotopies of the boundary curves can be extended to Whitney disks, and km is invariant under regular homotopy. Therefore km depends only on the pairing of intersection points, not on the choice of arcs.

A pairing of intersection points can be deformed to any other pairing by a sequence of moves, each of which changes two pairs. It is therefore sufficient to see that such a move does not change the invariant.

Suppose a_1, a_2 are positive and b_1, b_2 are negative intersection points between $h(D_i)$ and $h(D_j)$, all with the same associated element of the fundamental group. Let B_1, B_2 be (weak) Whitney disks for the pairs $\{a_1, b_1\}$ and $\{a_2, b_2\}$. Suppose C_1 is a Whitney disk for $\{a_1, b_2\}$. Then a Whitney disk C_2 for $\{a_2, b_1\}$ can be constructed from parallel copies of B_1, B_2, and C_1, and small twists near $\{a_1, b_2\}$.

C_2 has the same intersections as the union of the other disks, so $C_1 \cup C_2$ has the same intersections mod 2 as $B_1 \cup B_2$. Therefore km is invariant under replacement of $C_1 \cup C_2$ by $B_1 \cup B_2$.

This completes the proof of the lemma, and therefore of the existence part of the π_1-null embedding theorem. ∎

10.9 5-dimensional embeddings

The uniqueness part of the embedding theorem will be deduced from an analogous embedding theorem for 5-manifolds. Specifically, suppose $g\colon (V \times I, V \times \{0,1\} \cup \partial_0 V \times I) \to W \times I$ is a homotopy rel boundary between embeddings of V. Set $(X, \partial_0 X) = (V \times I, V \times \{0,1\} \cup \partial_0 V \times I)$, and $Y = W \times I$, then g is a map which is an embedding on $\partial_0 X$. A π_1-negligible embedding which agrees with g on $\partial_0 X$ is exactly the "concordance" required for the uniqueness statement.

Theorem. *Suppose* $(X; \partial_0 X, \partial_1 X)$ *is a compact 5-manifold triad with* $(X, \partial_0 X)$ *2-connected and* $(X, \partial_1 X)$ *1-connected. Suppose* $g\colon X^5 \to Y^5$ *is a map which embeds* $\partial_0 X$ *in* ∂Y, *and* $H_f^4(Y_g, X \cup \partial Y; \mathbf{Z}\pi_1 Y) = 0$. *Then there is an obstruction* $\tilde{\mu}d_g \in H^3(X, \partial_0 X; (\mathbf{Z}/2)[T_+])$ *which vanishes if and only if there is a* π_1-*negligible embedding* $g'\colon X \to Y$ *equal to* g *on* $\partial_0 X$ *and inducing the same homomorphism* $H_3(X, \partial_0 X; \mathbf{Z}\pi_1 Y) \to H_3(Y, \partial_0 X; \mathbf{Z}\pi_1 Y)$.

H_f denotes, as in 10.5, the finite cocycle cohomology, and Y_g the mapping cylinder of g. T_+ is the subset of $\pi_1 Y$ of elements of exponent exactly 2 on which ω_1 is trivial, as in 10.5. In particular if $\omega_1\colon \pi_1 Y \to \mathbf{Z}/2$ is injective on elements of order 2 then the obstruction group is trivial, and embeddings exist.

As explained above, the uniqueness in 10.5 follows by applying this to a homotopy $g\colon (V \times I, V \times \{0,1\} \cup \partial_0 V \times I) \to W \times I$. Delete a ball from each component of V which does not intersect $\partial_0 V$, to get the proper connectivity.

Proof: The proof is rather lengthy, so we begin with an outline. The first step is the definition of the obstruction, in terms of selfintersections in a 6-manifold. Then there is a reduction to the case where $(X, \partial_0 X)$ has a handlebody structure with only 3-handles. The images of these handles have 1-dimensional intersections and selfintersections. The invariant is described in terms of these intersections, and in fact only involves circles with connected preimages. Then the process of simplifying intersections begins, with the elimination of arcs joining singular points. Next is a version of the Whitney move, which joins intersection circles. When the invariant vanishes this is used to eliminate intersection circles with

connected preimage. This leaves circles of intersection whose preimages
are two circles. These are eliminated (after some preparation) by a
surgery procedure, which may change the homotopy class of the map.
The result is a map with no intersections or selfintersections among the
handles, so an embedding.

We begin by describing the obstruction. The connectivity hypotheses
imply that the relative homology and cohomology of $(X, \partial_0 X)$ is trivial
except in dimension 3. Therefore the obstruction group is given by
$H^3(X, \partial_0 X; (\mathbf{Z}/2)[T_+]) = \hom_\pi(H_3(X, \partial_0 X; \mathbf{Z}\pi), (\mathbf{Z}/2)[T_+])$, where π
denotes $\pi_1 Y$. The obstruction will be described as a homomorphism on
H_3.

The homomorphism $d_g \colon H_3(X, \partial_0 X) \rightarrow H_3(X \cup Y)$ is induced by
the chain map $1 - g_*$, as in 10.5. Multiplying by $\mathbf{R}$ takes this to
$H_3(X \cup Y \times \mathbf{R})$, where intersection and selfintersection forms are defined. The intersection form vanishes since homotopy classes can be
separated using the $\mathbf{R}$ coordinate. The selfintersection form takes val-
ues in $\mathbf{Z}\pi/((1+\bar\omega)\mathbf{Z}\pi)$. Recall that in an $2n$-manifold λ satisfies $\lambda(x, y) =
(-1)^n \bar\omega \lambda(y, x)$, and $\mu(x)$ is defined modulo $((1 + (-1)^{n+1}\bar\omega)\mathbf{Z}\pi)$. Before
now we have been working in dimension 4. Here the dimension is 6,
so the signs have changed. Except for the coefficient on the identity
element in $\mathbf{Z}\pi$ these satisfy $\lambda(x, x) = \mu(x) - \bar\omega\mu(x)$ (see 1.7), so the reduced selfintersection takes values in the kernel of this symmetrization
homomorphism. This kernel can be naturally identified with $(\mathbf{Z}/2)[T_+]$.

Assembling these gives $\tilde\mu(d_g \times \mathbf{R}) \colon H_3(X, \partial_0 X) \rightarrow (\mathbf{Z}/2)[T_+]$, which
will be the obstruction to embedding.

The first objective in the proof is to arrange $(X, \partial_0 X)$ to have a handlebody structure with only 3-handles. This is done in several steps,
each of which involves "stabilizing" the problem. To stabilize, suppose
g is an embedding on a collar of $\partial_0 X$, and consider the collar as a handlebody with a cancelling pair of 2- and 3-handles. Delete the 2-handle
from both X and Y to obtain $g' \colon X' \rightarrow Y'$. This map also satisfies the
connectivity and homological hypotheses of the theorem. If g' can be
replaced by an embedding agreeing with g' on $\partial_0 X'$, then an embedding
of X is obtained by replacing the 2-handle. It is therefore sufficient to
consider the stabilized problem.

X is changed by the addition of the 3-handle. Specifically X' is the
boundary connected sum $X \#_{\partial_0} S^2 \times D^3$. This new 3-handle is embedded
in both X' and Y', so the obstruction homomorphism vanishes on it.
This shows the obstruction for g' is the image of the obstruction for g.
$\partial_0 X$ is changed by connected sum with $S^2 \times S^2$, and $\partial_1 X$ is unchanged.
The homotopy type of X' is the 1-point union $X \vee S^2$.

Choose a handlebody structure on $(X, \partial_0 X)$ (one exists since X is a

5-manifold; see 9.1). According to the "geometrical connectivity" theorem (Wall [2]) a 1-connected 5-manifold pair has a handlebody structure without 0- and 1-handles. More precisely these handles can be
eliminated from a given structure at the expense of introduction of 2-
and 3-handles. Applying this to both $(X, \partial_0 X)$ and $(X, \partial_1 X)$ gives a
structure with only 2- and 3-handles. Basically we would like to make
g an embedding on the 2-handles, delete them as in stabilization, and
concentrate on the 3-handles. It is a bit difficult to get the 3-handles
disjoint from the 2-handles, and to relate the properties of the resulting g' to those of the original g, so we use a less direct approach. We
show that after a suitable number of stabilizations the resulting X' has
a handlebody structure with only 3-handles.

$(X, \partial_1 X)$ has a handle structure dual to the one on $(X, \partial_0 X)$; let Z denote $\partial_1 X$ union with the 2-handles in this dual structure. Obstructions
to factoring the identity of X (up to homotopy, rel $\partial_1 X$) through the
inclusion $i: Z \to X$ lie in $H^i(X, \partial_1 X; \pi_i(X, Z))$. These groups vanish,
since Z is a 2-skeleton if $i \leq 2$, and by duality using the 2-connectivity
of $(X, \partial_0 X)$, if $i > 2$. Therefore there is $j: X \to Z$, with ij homotopic
to the identity.

This factorization gives a decomposition $H_i(Z) = H_i(X) \oplus H_{i+1}(X, Z)$
($\mathbf{Z}\pi_1 X$ coefficients). But $H_{i+1}(X, Z)$ is concentrated in dimension 3,
and is free (generated by the 3-handles). $H_2(Z) = \pi_2(Z)$, so choose
maps $S^2 \to Z$ representing the generators of $H_3(X, Z)$. Together with
j these define $X \vee kS^2 \to Z$ which induces a simple equivalence of
chain complexes with $\mathbf{Z}\pi_1 X$ coefficients. Since both complexes have
fundamental group $\pi_1 X$, the map is a simple homotopy equivalence.

The conclusion is that the manifold X' obtained by k-fold stabilization is simple homotopy equivalent rel $\partial_1 X$ to a handlebody Z (on $\partial_1 X$)
with only 2-handles. Approximate the equivalence rel $\partial_1 X$ by an embedding $Z \to X$. This is be done using the 5-dimensional analog of the
immersion lemma, and might involve changing the attaching maps of
the 2-handles by rotations. (Although it is easy to see that for the particular Z constructed above no rotations are necessary.) Let W denote
the closure of the complement of Z in X.

W is an s-cobordism from $\partial_0 X$ to the rest of its boundary. The
fact that the inclusion is a simple equivalence implies by excision and
duality that W is a simple $\mathbf{Z}\pi_1 X$ coefficient homology H-cobordism,
so it is sufficient to see the boundary pieces and W all have the same
fundamental group as X. $\partial_0 X$ does by the 2-connected hypothesis. W
does since it is obtained by deleting 2-handles from X, and X is 5-
dimensional. Similarly $\partial_1 W$ is obtained by deleting 2-handles from Z,
so has the same fundamental group as Z.

Further stabilization of the problem can be arranged to change W by connected sum along an arc with $S^2 \times S^2 \times I$, as in the exercise preceeding 7.3. To see this choose the cancelling 2- and 3-handle pair to lie in a neighborhood of a collar arc going from $\partial_0 X$ to $\partial_1 X$ (missing the handles). Delete a neighborhood of the 2-handle to get X', and add the dual of the 3-handle to Z. Then the dual also gets deleted to form W', leaving the desired arc sum.

Finally apply the stable 5-dimensional s-cobordism theorem (in the exercise preceeding 7.3) to conclude that after some number of stabilizations the associated manifold W is a product. This means X is homeomorphic to Z, so $(X, \partial_1 X)$ has only 2-handles. Dually $(X, \partial_0 X)$ has only 3-handles, as desired.

Now we construct "transverse spheres" for the images of the 3-handles in Y. If the core disks of the handles are denoted D_i, then the objective is a collection of embeddings $\alpha_i \colon S^2 \to Y$ so that $g(D_i) \cap \alpha_j$ is empty unless $i = j$, and then is a single (transverse) point. These embeddings are not required to be framed.

As in the proof of 10.5(1) in 10.8 the vanishing of the cohomology group $H_f^4(Y_g, X \cup \partial Y; \mathbf{Z}\pi_1 Y)$ implies there are elements $\alpha_i \in \pi_2(Y)$ with intersection numbers $\lambda(g(D_i), \alpha_j) = 0$ if $i \neq j$, and $= 1$ if $i = j$. Since $\dim(Y) = 5$ these can be approximated by embeddings. The extra intersections between these and the $g(D_i)$ can be arranged in pairs with Whitney disks. Again since $\dim(Y) = 5$ the Whitney disks can be chosen to be embedded, although their interiors may intersect the $g(D_i)$. Push the $g(D_i)$ across these Whitney disks. This may introduce new intersections among the $g(D_i)$, but reduces intersections with the α_j to single points.

The rest of the proof involves studying the intersections among the images of the 3-disks $g(D_i)$. Approximate them to be in general position and transverse to each other, then the intersections are circles and arcs of double points. These are disjoint from the boundaries of the disks since g is an embedding there. g is an immersion except at isolated "cusps" ocurring at the ends of arcs of double points. The preimage in $\cup D_i$ is a union of circles. Over a circle of intersections the preimage is a 2-fold covering space, so either two circles each going by homeomorphism, or a single circle going around twice. There is a single circle over an intersection arc, a 2-fold "branched cover" branched over the endpoints.

We describe the obstruction in terms of these intersections. For this it is sufficient to determine the value of $\tilde{\mu}(d_g \times \mathbf{R})$ on generators; the classes represented by the handle cores D_j. The images in $H_3(X \cup Y)$ are represented by $D_j \cup g(D_j)$. Let $f_j \colon S^3 \to \mathbf{R}$, then the graphs of these in $(X \cup Y) \times \mathbf{R}$ represent $(d_g \times \mathbf{R})([D_j])$.

Intersections among these occur at points identified both by g and f_j. Consider a circle of intersections of g which is covered by two circles in D_*. The function f can be arranged to have values on one component strictly greater than on the other, so this circle produces no intersections in $(X \cup Y) \times \mathbf{R}$. The intermediate value theorem implies that arcs, and circles covered by a single circle, must have points for which the two points in the preimage are taken to the same value by f. But f can be arranged so there is exactly one such point on each such arc or circle, so these produce exactly one selfintersection point each. $\tilde{\mu} \in (\mathbf{Z}/2)[T_+])$ therefore counts these mod 2; if $\gamma \in T_+$ then the coefficient on γ in $\tilde{\mu}(D_j \cup g(D_j))$ is the number mod 2 of intersection circles covered by a single circle, and with associated π_1 element γ. The arcs do not show up since they have associated element $1 \in \pi$, and 1 is omitted from $\tilde{\mu}$ and T_+.

A selfintersection circle covered by a single circle is itself a representative for its associated element in $\pi_1 Y$. Since the double cover is nullhomotopic, it follows that it is an element of exponent 2. The fact that D is self-transverse at this circle divides the normal bundle of the circle locally into two 2-dimensional subbundles. The double cover splits globally as a sum of orientable subbundles. This identifies the normal bundle of the circle in Y as being obtained from the bundle $\mathbf{R}^2 \times \mathbf{R}^2 \times I$ over I by identifying the ends via the matrix $\left[\begin{smallmatrix} 0 & I \\ I & 0 \end{smallmatrix}\right]$ This is orientable, so as a π_1 element it lies in the kernel of ω_1. Therefore the elements which can occur this way are $T_+ \cup \{1\}$. We conclude that the obstruction $\tilde{\mu}(d_g \times \mathbf{R})$ vanishes if and only if: for each j and $\gamma \neq 1$ in $\pi_1 Y$ there are an even number of selfintersection circles of $g(D_j)$ covered by a single circle and with associated group element γ.

We begin the modification of intersections by first eliminating arcs. The map has no arcs if and only it is an immersion, so the fastest way to do this is to use the immersion theorem and the fact that $\pi_3(B_{TOP}) = 0$. A geometric argument will be given because it gives additional information about circles double covered by a circle.

At each end of an arc of intersections is a "cusp" as pictured in 1.6 (but multiplied by $(\mathbf{R}^2, \mathbf{R})$ to make it 5-dimensional; see Whitney [2]). The basic idea is that the cusps on the ends of an arc can be joined together to give an intersection circle. After this is done the map is an immersion.

In detail, a small ball about the singular point intersects the image $g(D_j)$ in the cone on a standard immersion of S^2 in S^4, with a single selfintersection. This immersion is obtained from a 2-sphere in 3-space with a line of intersections by pushing one sheet into the future on one side of the midpoint of this line, and into the past on the other side.

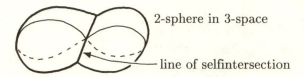

2-sphere in 3-space

line of selfintersection

There are two of these, depending on which side of the midpoint is pushed into the future, and these are distinguished by the sign of the intersection point. In a 5-manifold these signs depend—as do intersection numbers—on choices of paths from the handle images to the basepoint, and an orientation at the basepoint. Transport the orientation along the path to get an orientation at the singular point, and use the induced orientation on the boundary of a ball about that point. Cusps on opposite ends of an arc of intersections have opposite signs.

Cusps of opposite sign can be "cancelled." Choose an arc between them disjoint from the double curves. A neighborhood of this arc can be thought of as a boundary connected sum of ball neighborhoods of the endpoints. The intersection of the handle image with the boundary of this neighborhood is the connected sum of two standard immersed spheres. If the intersection points have opposite sign then this is also the boundary of the product with I of a standard immersion of D^2 in D^4 with a single selfintersection. Replacing the original intersection with this product removes the cusps and joins the ends of the original double point arcs. As observed above joining the ends of all double arcs in this way gives an immersion with only double circles.

When ends of a double arc are joined in this way there is a choice which determines whether the resulting circle is covered by a single circle or by two circles. As above consider the intersection of the handle image with the boundary of the arc neighborhood as a sum of two immersed 2-spheres. The resulting sphere has two intersections of opposite sign, and there are two Whitney disks, differing by a twist, which can be used to remove the intersections leaving a standard (unknotted) 2-sphere.

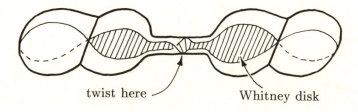

twist here   Whitney disk

Put the trace of the Whitney move in a collar of the boundary S^4, and then fill in with a standard embedded $D^3 \subset D^5$. The track of the intersection points in the Whitney move is the new double arc. By following sheets near this arc the structure of the double cover can be deduced, and in particular single twists in the sum tube switch the double cover between a single circle and two circles.

The intersection points in circles generated from arcs have trivial associated element in the fundamental group. This is because near the singularity there are loops going from one sheet to the other inside a ball. The discussion above implies that we can introduce an arbitrary number of circles doubly covered by a single circle, and with trivial associated element in π_1. To do this introduce folds with arcs of double points, then join the ends in such a way to get a connected double cover.

We now describe a version of the Whitney isotopy, which will be used to eliminate circles with connected double covers.

Suppose x and y are points of intersection between $g(D_i)$ and $g(D_j)$, whose associated group elements are equal. Join the preimages by embedded arcs in both D_i and D_j, then the resulting loop in Y is contractible. Approximate a nullhomotopy by an embedded disk. The interior of this disk may have point intersections with the $g(D_*)$, which can be removed using connected sums with the transverse spheres α_*. A neighborhood of this disk can be parameterized by the model shown below, modified to be 5-dimensional; multiply by $I \times I$, and multiply D_i, D_j by the first and second coordinates respectively.

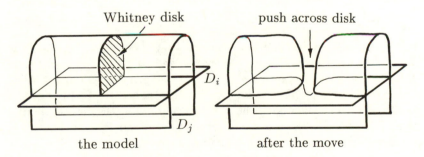

Whitney disk   push across disk

D_i

D_j

the model   after the move

The reason this is always possible, and there is not an "opposite sign" condition as with isolated intersection points is that there is an additional degree of freedom: the orientations of the intersection curves can be changed. The picture on the right shows D_i, D_j after a Whitney isotopy across the disk. The two intersections circles are modified by a connected sum operation at the points x and y.

For the first application of this Whitney move we observe that if x

and y lie on different circles, both covered by a single circle, then the operation combines the circles to give a single circle covered by two circles. This implies that if there are an even number of circles with connected double cover and some fixed associated $\gamma \in \pi_1$ then they can be joined in pairs and all converted to trivially covered circles.

We conclude from this that if the obstruction $\tilde{\mu}(d_g)$ vanishes g can be modified to have no circles with connected double cover. First, vanishing means that for any nontrivial element in $\pi_1 Y$ there are an even number of circles in each D_j with connected cover and that particular element. Thus all these can be combined, leaving only circles with trivial associated π_1 element (or disconnected cover). If the number of these in some D_j is not even then according to the analysis of arcs we can introduce a single new one, and then eliminate these also.

Now we describe a type of "surgery" on $g(D_*)$, which will be the main technique for removing intersection circles. Suppose a circle of intersections has preimage two circles $r \subset D_i$ and $r' \subset D_j$. Begin with the simplest case where r is unknotted and unlinked, so suppose $C \subset D_i$ is an embedded 2-disk with boundary r and interior disjoint from the double points of g. This data will be used to construct a 4-dimensional handlebody with D_j at one end, and a map into Y extending g.

Start with $D_j \times I$, and add a 2-handle on $r' \times \{1\} \subset D_j \times \{1\}$. Map this to the normal disk bundle of the image of D_i, restricted to the 2-disk C (note that requiring this map to be defined singles out a specific framing of $r' \times \{1\}$ on which to attach the handle). To construct a second 2-handle begin with the core of the dual of the first handle (a fiber of the disk bundle over C). Perturb the interior to be disjoint from the construction to date. There is a single intersection point of this disk with D_i, remove this by connected sum with a copy of the transverse 2-sphere α_i. This gives an embedded 2-disk in Y. To get a 2-handle we need a 2-disk bundle over this extending the normal bundle of the boundary circle in the boundary of the handlebody. The normal bundle in Y is 3-dimensional, and the complement of the 2-dimensional bundle over the boundary is trivial, so according to the stability of bundles it is possible to extend the splitting over the 2-disk. This splitting is not unique, and we discuss the effect of different choices below.

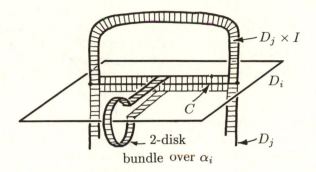

The handlebody gives a bordism from g on D_j to a map on some 3-manifold, which has the same selfintersections and intersections with the other D_* as D_j, except that the image of $r \cup r'$ has been eliminated. The bordism shows that this manifold and D_j represent the same relative homology class, so we can replace D_j with it (thereby simplifying the situation) provided the new manifold is a 3-disk. We describe two situations in which this is the case.

First, if the sphere α_i can be framed then the manifold can be arranged to be a disk. If α_i can be framed (which is equivalent to $\omega_2(\alpha_i) = 0$) then there is a trivial 2-disk subbundle of its normal bundle. Connect sum with the trivial bundle does not change framings on the boundary, so in this case the second 2-handle is attached on the same framing of the boundary S^1 as the framing coming from the dual disk of the first 2-handle. The effect on the boundary is the same as deleting a neighborhood of the dual disk. However this operation undoes the first handle attachment, and gives D^3.

For the other case suppose that the circle $r' \subset D_j$ bounds an embedded disk (not necessarily disjoint from the other double points). This disk defines a trivialization of the normal bundle, which differs by rotations (ie. by an element of $\pi_1(O(2)) \simeq \mathbf{Z}$) from the framing used to attach the first 2-handle. If these framings agree then the new manifold is a 3-disk. This is because the disk and the core of the first 2-handle fit together to give an embedding of $S^2 \times D^2$ in the handlebody. Replacing this with $D^3 \times S^1$ gives a new handlebody with a 1-handle and a 2-handle (the second 2-handle in the original). The 2-handle is attached on a circle which intersects the dual sphere of the 1-handle in exactly one point, so the two handles can be cancelled. Again the modified handlebody is a collar on D_j and the complementary piece of the boundary is a 3-disk.

The next step is to obtain embedded disks bounding preimage cir-

cles, so the surgery procedure above can be applied. To begin with
there are immersed disks bounding the circles, whose intersections are
all "clasps." To get these choose nullhomotopies in general position, so
immersed except at isolated cusps. Push cusps across the boundary (this
introduces twists about the boundary circle which change the framing).
If there are circles of intersection, push one disk across the edge of the
other to convert them to arcs. Then triple points can be pushed along
a double arc across the boundary, and eliminated. Finally if there is an
arc lying entirely in the interior of a disk, a piece of its interior can be
pushed across the boundary to convert it into two clasps. None of these
modifications involve changing the boundary circles.

We now associate a "rotation number" to preimage circles. If $r \subset
D_i$ is the preimage of $g(D_j)$ then the normal bundle of r in D_i is (by
transversality) the restriction of the normal bundle of $g(D_j)$ in Y. The
contractibility of D_j defines a trivialization of this bundle. But r is also
the boundary of an immersed disk in D_i, which also gives a trivialization
of the bundle. These trivializations differ by a rotation in $\pi_1(O(2)) \simeq \mathbf{Z}$,
and this is defined to be the rotation number of r.

Now clasps will be removed. Suppose C_1 and C_2 are disks in D_i with
boundaries preimages of D_j and D_k respectively. Suppose these intersect
in a clasp arc. A finger move of D_j along this arc gives an immersion in
which the clasp is removed, but introduces a new circle of intersections
between D_j and D_k. There are standard embedded 2-disks, say E and
E', bounding the preimage circles, and there are clasps between these
and disks in D_j and D_k.

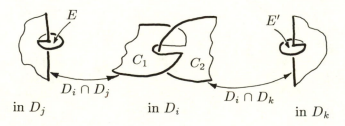

$$\text{in } D_j \qquad D_i \cap D_j \qquad \text{in } D_i \qquad D_i \cap D_k \qquad \text{in } D_k$$

These new intersection circles have rotation number 0 since they arise
from finger moves. We repeat the move, pushing D_i along the clasp in E.
This introduces new D_i, D_k intersections with disks F, F' which clasp
E' and C_1 respectively. But E is now embedded disjointly from other
double points, and has rotation number 0, so the surgery procedure can
be used to change the immersion to eliminate it. The disk E' is also
removed, so after this the disk F is disjointly embedded. It also has
rotation number 0, so can be removed by surgery. After doing this we

have an immersion with all the same data as the original except that the clasp in $C_1 \cap C_2$ has been removed.

Applying this operation at each clasp yields an immersion whose preimage circles bound disjointly embedded disks.

At this point we can use the surgery procedure to eliminate a great many circles: any (trivially covered) circle involving a D_j with $\omega_2(\alpha_j) = 0$, or with trivial rotation number. In fact we can complete the proof of the theorem if ω_2 does not vanish on the subspace of $\pi_2 Y$ perpendicular to the image of $H_3(X, \partial_0 X; \mathbf{Z}\pi_1 Y)$ in a sense analogous to the condition in 10.5. This condition is equivalent to the existence of a 2-sphere disjoint from the image of g, with nontrivial ω_2. Using connected sums with this sphere we can arrange that all α_j can be framed. This means all trivially covered circles can be eliminated. Since vanishing of the $\tilde{\mu}(d_g)$ obstruction implies the nontrivially covered circles can be eliminated, we can get an embedding.

To complete the proof in the remaining cases we show how to change the rotation numbers of circles in D_i when $\omega_2(\alpha_i) \neq 0$. If g can be modified so all the rotation numbers are zero, then circles can be eliminated using the surgery procedure independently of $\omega_2(\alpha_*)$.

Suppose $\omega_2(\alpha_i) \neq 0$. Then the normal bundle of α_i in Y is the sum of the 2-dimensional bundle with Euler number 1 (see 1.7) and a trivial line bundle. The sphere bundle of this bundle is the Hopf fibration $S^3 \to S^2$. Let $E \colon S^3 \to Y$ denote the embedding of this sphere bundle, then $E \cap g(D_i)$ is a circle; a fiber of the Hopf fibration.

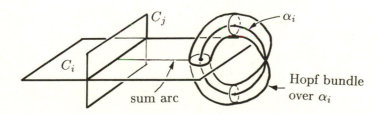

The preimages of this circle bound embedded disks in both S^3 and D_i. The rotation number of the circle in S^3 is 1: the preimage of the normal bundle of $g(D_i)$ is the same as the preimage under the Hopf map of the normal bundle of a point in S^2. That this bundle has rotation number 1 can be seen from the fact that distinct fibers of the Hopf map have linking number 1: linking numbers are defined in terms of intersections with bounding disks, so in the case of images of standard sections of a framing linking is the same as the rotation number.

Now suppose $x \in g(D_i) \cap g(D_j)$. Choose an arc in D_i from the preimage of x to a point in $E \cap g(D_i)$. Denote by g' the map of $D_j = D_j \# S^3$ obtained by connected sum of $g(D_j)$ with E. This map induces the same homomorphism on homology as g because E bounds the disk bundle, so is trivial in $H_3(Y)$. All intersection structure is the same except that the circle containing x is replaced by its connected sum with $E \cap g(D_i)$. Rotation numbers are additive under connected sum, so the rotation number of the new circle is greater by 1 than that of the old. Alternatively, reversing the orientation of E gives a sphere with intersection circle with rotation number -1, and connected sum with this reduces rotation numbers by 1.

All rotation numbers can be changed to be 0 by adding such spheres repeatedly. This gives intersection circles which can be eliminated by surgery, so completes the proof of 10.9. ∎

10.10 References

The classification theorem 10.1 comes from Freedman [2]. The characterization of connected sums in 10.3, the definition of $*W$ in 10.4, and the π_1-negligible embedding theorem of 10.5 all appear here for the first time.

The characterization theorem has been extended in several ways. Simply connected 4-manifolds with boundary have been described by Vogel [1], and Boyer [1]. The surgery approach to classification has been pushed through for certain classes of finite fundamental groups by Hambleton and Kreck [1]. The results are quite a bit more complicated than the trivial and infinite cyclic cases treated here.

CHAPTER 11

Surgery

"Surgery" refers to a collection of techniques for manipulating manifolds. The brief conclusion is that "surgery works" for topological 4-manifolds, provided the fundamental groups are good. In this chapter we survey some of the main developments with interesting 4-dimensional applications, but do not attempt a comprehensive treatment.

The chapter begins with the "plus construction," which illustrates the basic technique without a lot of machinery. This is used to construct group actions on contractible 4-manifolds in 11.1C. The surgery sequence is described and the new cases of exactness are proved in 11.3. The "surgery program" for classifying manifolds using this sequence is also discussed. This program is applied to the classification of simply-connected manifolds in 11.4 and certain aspherical manifolds in 11.5.

The surgery program is worked out in detail (modulo "goodness" problems with fundamental groups) for 4-manifolds homotopy equivalent to 1-complexes. Section 11.7 concerns knots, giving characterizations of unknotted 2-spheres in the 4-sphere and $\mathbf{Z}$-slice knots in homology 3-spheres. There is a version of surgery which gives homology equivalence with general coefficients. This does not extend to dimension 4, and this failure is discussed in 11.8. Section 11.9 gives a weak version of the end theorem for tame ends of 4-manifolds, and describes the use of surgery to classify tame ends. Finally in section 11.10 we consider the classification of neighborhoods of fixed points of group actions on 4-manifolds.

11.1 The plus construction

This is adapted from a homotopy theory technique. Versions for CW complexes and higher dimensional manifolds will be described in the next section.

11.1A Theorem. *Suppose M is a compact 4-manifold, π a finitely presented good group, and $f_*: \pi_1 M \to \pi$ is onto and has perfect kernel. Then there is a compact bordism rel boundary W from M to a manifold M^+ so that $\pi_1 W = \pi$, the inclusion $M \to W$ is a simple $\mathbf{Z}\pi$ homology equivalence, and $M^+ \to W$ is a simple homotopy equivalence. These properties specify W uniquely up to homeomorphism rel M.*

W is a bordism "rel boundary" if $\partial W = M \cup M^+ \cup \partial M \times I$. Note W is a simple $\mathbf{Z}\pi$-homology s-cobordism.

Proof: Denote the kernel of $f_* \colon \pi_1 M \to \pi$ by Δ. Since f_* is onto and the two groups are finitely presented, Δ is finitely normally generated. This means there is a finite set $g_1, \ldots, g_n \in \Delta$ so that Δ is the smallest normal subgroup of $\pi_1 M$ containing them. Let $\hat{M}$ be the cover of M with fundamental group Δ. Choose lifts $\hat{g}_* \colon S^1 \to \hat{M}$. The homology $H_1(\hat{M}; \mathbf{Z})$ is the abelianization of Δ, so is trivial since Δ was presumed to be perfect. Therefore the $\hat{g}_*$ extend to maps of compact connected oriented surfaces bounding the circles; $\hat{G}_i \colon A_i \to \hat{M}$. Let G_* denote the compositions into M.

The maps G_* are homotopic to disjoint framed embeddings. The immersion lemma gives immersions with isolated intersections, which can then be removed by finger moves of one surface across the boundary of the other. The normal bundle can be framed because it is determined by ω_1, which must vanish because $H_1(\hat{M}; \mathbf{Z}/2) = 0$.

Construct a 5-dimensional handlebody W_1 by attaching 2-handles to the framed curves $g_* \times \{1\} \subset M \times I$. Denote the closure of the complement of $M \times \{0\}$ in ∂W_1 by M_1. The fundamental group of W_1 is that of M divided by the normal closure of the g_*, so is $\pi_1 M/\Delta = \pi$. W_1 is obtained from M' by adding 3-handles, which do not change π_1, so $\pi_1 M_1$ is also π.

There are disjoint framed embedded closed surfaces $G_i \cup D_i$ in M_1 formed by adding copies of the cores of the handles to the surfaces G_i. These surfaces have disjointly embedded transverse spheres, namely the boundary spheres of the dual handles. Finally we observe that the $G_i \cup D_i$ are homologous to maps of spheres. This is because they have been constructed to lift to the cover corresponding to the homomorphism $\pi_1 M_1 \to \pi$, which is the universal cover.

Since π has been assumed to be good, corollary 5.1B applies to the homology classes represented by the $G_i \cup D_i$. We conclude that they are represented by framed π_1-negligibly embedded 2-spheres in M_1. Define W to be the handlebody obtained by attaching 3-handles to W_1 on these framed embeddings.

This W satisfies the conclusions of the theorem. The fundamental group is still π because adding 3-handles does not change π_1. Denote by M^+ the closure of the complement of $M \cup \partial M \times I$ in ∂W, then this also has $\pi_1 M^+ = \pi$, because of the π_1-negligibility of the embeddings. The pair (W, M) is a handlebody with only 2- and 3-handles, with boundary homomorphism in the $\mathbf{Z}\pi$ chain complex given by the identity matrix. Therefore the inclusion $M \to W$ is a simple $\mathbf{Z}\pi$-coefficient homology isomorphism, and by duality so is $M^+ \to W$.

To complete the proof we show that W is unique. Suppose W' is another such bordism. Consider $V = (W \cup_M W') \times I$ as an s-cobordism with boundary, from $W \times \{0\}$ to $W' \times \{0\}$. The boundary s-cobordism is $(W \cup_M W') \times \{1\}$, which is 5-dimensional with good fundamental group. According to 7.1 this is homeomorphic to $M^+ \times I$. Then according to the higher-dimensional s-cobordism theorem this product structure extends to a product structure on V. In particular there is a homeomorphism $W \simeq W'$, which is the identity on M. ∎

Exercise (1) Show that if M is a 4-manifold (possibly noncompact), $\Delta \subset \pi_1 M$ is a normally finitely generated perfect subgroup and $\pi_1 M / \Delta$ is good, then there is a compactly supported $\mathbf{Z}[\pi_1 M / \Delta]$ homology s-cobordism rel boundary W from M to M^+ such that $\pi_1 M^+ = \pi_1 W = \pi_1 M / \Delta$. (A bordism is "compactly supported" if it is a product in the complement of a compact set disjoint from ∂M.) Show also that such a cobordism is unique up to homeomorphism fixing M and some $U \times I$, where $U \times I$ is the given product structure and $W - U \times I$ is compact. ☐

Corollary. *A 4-manifold M bounds a compact contractible (topological) 5-manifold if and only if it is a closed $\mathbf{Z}$ homology sphere.*

Proof: This is the 4-dimensional analog of corollary 9.3C. The only if direction is a straightforward homology calculation. For the other direction note that $\pi_1 M \to \{1\}$ has perfect kernel. The trivial group is good, so the theorem applies to give a 1-connected homology h-cobordism W to a 1-connected manifold M^+ with the same homology. According to the Poincaré conjecture (7.1B) M^+ is the 4-sphere, so $W \cup_{M'} \mathrm{cone} M'$ is the required contractible manifold. ∎

There are proofs of this which do not use the 4-dimensional embedding theorem, so in a sense are more elementary. This proof also works in higher dimensions.

A special case of the plus construction was used in the proof of 9.3C to show that a 3-dimensional homology sphere bounds a contractible manifold. The next result is an equivariant version of this.

11.1C Proposition. *A free action of a finite group on a 3-dimensional homology sphere extends, uniquely up to equivariant homeomorphism, to an action on a contractible manifold which is free except at one fixed point.*

Proof: Suppose G acts freely on the homology sphere M. The kernel of $\pi_1(M/G) \to G$ is $\pi_1 M$, so is perfect. Since finite groups are good, theorem 11.1A applies to $(M/G) \times [i, i+1]$ to give bordisms W_i (rel boundary) to manifolds N_i with fundamental groups G. G then acts on the universal cover $\tilde{N}_i$. Let V be the 1-point compactification of $\bigcup_{i\geq0} \tilde{N}_i$, then as in 9.3C V is a contractible manifold with boundary M. The covering transformations give an action of G on V which is free except at the compactification point, so this is an extension of the type required for the proposition.

For the uniqueness, suppose that U is a contractible manifold with boundary M, on which G acts fixing only a point p. We can regard $(U - p)/G \times [0,\infty)$ as a proper $\mathbf{Z}G$ homology s-cobordism from $(U - p)/G \times \{0\}$ to $M \times [0,\infty)$. The union of the bordisms constructed above, $\bigcup_{i\geq0} W_i$ is a proper homology s-cobordism from $M \times [0,\infty)$ to $\bigcup_{i\geq0} N_i$. Putting these together gives a $\mathbf{Z}G$ homology h-cobordism from $(U-p)/G \times \{0\}$ to $\bigcup_{i\geq0} N_i$. But this also has stable fundamental group G in a neighborhood of the end, so is a proper s-cobordism. According to the proper h-cobordism theorem 7.3 this has a product structure. In particular there is a homeomorphism rel M from $(U - p)/G$ to $\bigcup_{i\geq0} N_i$. Taking universal covers and 1-point compactifications gives an equivariant homeomorphism from U to the manifold V constructed above, so V is unique. ∎

Exercise Extend 11.1C to some actions which are not free: call an action on a contractible manifold *weakly conelike* if there is an open invariant subset $U \subset \partial M$ which extends to an equivariant embedding cone$U \to M$ such that G acts freely on the complement. (1) Suppose there is a PL action of G on a homology sphere N which is free in the complement of a subcomplex X, so that $\pi_1(N - X, x)$ has perfect commutator subgroup, for each basepoint x. Show the action extends uniquely up to equivariant homeomorphism to a weakly conelike action on a contractible manifold. (See the exercise after 11.1B)

The next exercise develops examples of X as above. Non-free points of linear actions are unions of linear subspaces. Therefore actions on S^3 coming from orthogonal actions on $\mathbf{R}^4$ are unions of "linear spheres," these being intersections of S^3 with linear subspaces. (2) Show that a union of a finite collection of linear subspheres in S^3 satisfies the π_1 condition in the exercise above, if there is at most one 1-sphere, or two disjoint 1-spheres and no 2-spheres. (This last is the Hopf link.) The result applies to situations homologically similar to these, for example the $\mathbf{Z}$-slice circles in homology spheres considered in 11.7B. (3) Show that if a locally flat $S^j \subset N$ satisfies the condition in (1) then the resulting embedding of cone$S^j = D^{j+1}$ in the contractible manifold is locally flat. (See 9.3A.) □

11.2 Plus in other categories and dimensions

For manifolds of dimension 5 or greater the statement of 11.1 is true, without the restriction that $\pi_1 X$ be good. The proof is the same except that the 2-spheres on which the 3-handles are attached are embedded simply by general position, not with any sophisticated theorem. The theorem is false for 3-manifolds; we observe in 11.9B that there are many homology lens spaces not homology h-cobordant to homotopy lens spaces.

The version for CW complexes gives a way to kill subgroups of the fundamental group, and is due to Quillen [1].

Proposition. *Suppose that X is a CW complex, and $\pi_1 X \to \pi$ is onto and has perfect kernel. Then there is a map $f \colon X \to Y$, unique up to homotopy equivalence, which on π_1 is $\pi_1 X \to \pi$, and is a $\mathbf{Z}[\pi]$ homology equivalence.*

Proof: Let Δ denote the kernel. Attach 2-cells to X on a set of generators of Δ and denote the result by X_1. Homologically ($\mathbf{Z}\pi$ coefficients) the attaching maps are trivial, so X_1 looks homologically like a 1-point union of X and some 2-spheres. Since $\pi_1 X_1 = \pi$, $H_2(X_1; \mathbf{Z}\pi) = \pi_2 X_1$. Therefore the new homology classes are represented by maps of 2-spheres, and we denote by X^+ the space obtained by attaching 3-cells to X_1 by these maps. The result is a map $X \to X^+$ which on π_1 is the quotient $\pi_1 X \to \pi$, and is a $\mathbf{Z}[\pi]$ homology equivalence.

This construction is well-defined up to homotopy equivalence. If $X \to Y$ is any map inducing the quotient on π_1 and an isomorphism on $\mathbf{Z}[\pi]$ homology, then there is a canonical homotopy class of homotopy equivalences $X^+ \to Y$ which homotopy commute with the given maps. To see this use the images of the attaching maps in Y to construct Y^+ and a commutative diagram

$$
\begin{array}{ccc}
X & \longrightarrow & Y \\
\downarrow & & \downarrow \\
X^+ & \longrightarrow & Y^+
\end{array}
$$

The maps $X^+ \to Y^+ \leftarrow Y$ are all isomorphisms on $\pi_1 = \pi$, and $\mathbf{Z}[\pi]$ homology. The Whitehead and Hurewicz theorems imply they are homotopy equivalences. A homotopy inverse for the second map gives a homotopy equivalence $X^+ \to Y$ as required. ∎

11.3 The surgery sequence

A central result of surgery is an exact sequence describing a set of homology equivalences of manifolds. If X is a space, the *structure set*

$S^h_{\text{TOP}}(X)$ is defined to be the set of equivalence classes of $\mathbf{Z}\pi_1 X$ homology equivalences $M \to X$, where M is a closed topological manifold. Two such are equivalent if there is a $\mathbf{Z}\pi_1 X$ homology h-cobordism W between them, with a map $W \to X$ extending the ones given on the boundary.

In order for $S^h_{\text{TOP}}(X)$ to be nonempty it is at least necessary that X be a Poincaré space (satisfy Poincaré duality with $\mathbf{Z}[\pi_1 X]$ coefficients), and we will generally assume this. (See Wall [1, §2], Browder [1].)

This set is usually defined using homotopy, not just homology, equivalences. If the dimension is at least 5, or is 4 and $\pi_1 X$ is good, then the plus construction shows that homology and homotopy equivalence give the same set of equivalence classes. We use homology equivalence here because the statement in 11.3A remains true in dimension 3 (and below) with this definition.

Variations: If X is a finite complex then the "simple" structure set $S^s_{\text{TOP}}(X)$ has objects simple homology equivalences, and s-cobordisms in the equivalence relation. If $N \subset X$ is a closed manifold then $S_{\text{TOP}}(X, N)$ has objects $M \to X$ with M compact and with boundary taken by homeomorphism to N. Other variations permit objects with boundary which is not held fixed and PL or differentiable manifolds. Homology equivalence with respect to other coefficients is discussed in 11.6D, and a noncompact version is used in 11.7C.

11.3A Theorem. *Suppose (X, N) is a Poincaré pair, with N a closed manifold. Then the surgery sequence*

$$L^h_{n+1}(\pi_1 X, \omega) \to S^h_{\text{TOP}}(X, N) \to NM_{\text{TOP}}(X, N) \to L^h_n(\pi_1 X, \omega)$$

is "exact," provided $\pi_1 X$ is good if $n = 4$.

The new case is the 4-dimensional one. If X is a finite complex then there is a version for simple structures obtained by replacing L^h and S^h with L^s and S^s. The ω in the notation is the first Stiefel-Whitney class $\omega_1 \colon \pi_1 X \to \mathbf{Z}/2$.

We begin with a formal proof, consisting of modifications required to make the high-dimensional proof work. The high dimensional proof is then sketched, for the benefit of readers unfamiliar with it. Explanations of the terms used in the theorem, and the meaning of "exactness," are given in 11.3B.

Proof: The proof in higher dimensions is given in Wall [1, Theorems 10.3 and 10.5]. The proof given there also works when $n = 3$, but was not mentioned because it produces $\mathbf{Z}\pi_1 X$-homology equivalences rather than homotopy equivalences. This is satisfactory and useful to us here. In dimension 2 the sequence is essentially trivial. Exactness is an easy consequence of the result of Eckmann and Linnell [1] that a 2-dimensional Poincaré space is homotopy equivalent to a manifold, together with the calculations of the surgery groups.

The 4-dimensional case is reduced in Wall [1, §5] to finding embedded 2-spheres representing a hyperbolic basis for a summand of π_2 of a 4-manifold. Explicitly this means: if A_i, B_i are framed immersed 2-spheres in a 4-manifold, $1 \le i \le n$, and all intersection and selfintersection numbers among them are zero except $\lambda(A_i, B_i) = 1$, then the A_i are regularly homotopic to disjoint embeddings. If the fundamental group is good, and balls intersecting the A_i are deleted to leave immersed 2-disks with algebraic transverse spheres B_i, then corollary 5.1B applies to give regular homotopies to the desired embeddings. ∎

Note that the conclusion is unaffected by changing the manifold up to s-cobordism. Therefore a solution to the embedding problem up to s-cobordism would imply validity of the surgery sequence for all fundamental groups. In the next chapter we will see these two problems are equivalent.

The proof given describes how to fill the gap left in Wall's procedure when $n = 4$ and fundamental groups are good. We briefly outline this procedure for even-dimensional manifolds, to explain how the embedding result is used. Suppose $n = 2k$. Begin with a normal map $M \to X$, and construct bordisms of this by attaching handles to $M \times I$. If $i < k$ then handles of index $\le i+1$ can be used to construct a bordism to $M' \to X$ which is $i + 1$-connected. If $M' \to X$ can be made $k + 1$-connected then it is a homotopy equivalence, by Poincaré duality. The tricky step is therefore going from k-connectivity to $k + 1$-connectivity, which requires addition of $k+1$-handles to the bordism. These handles are to be added to framed k-spheres embedded in an $2k$-manifold. Two problems occur; locating a collection of homotopy classes with the property that attaching handles to them will yield a homotopy equivalence, and then representing these homotopy classes by embeddings.

Wall shows that if the intersection and selfintersection forms on the relative homology $H_{k+1}(X, M)$ are "hyperbolic," then attaching handles on half the basis will give an equivalence. Hyperbolic means there is a basis in which the intersection form has matrix $\left[\begin{smallmatrix} 0 & I \\ (-1)^k I & 0 \end{smallmatrix}\right]$, and the selfintersection form vanishes on the basis. The group L_{2k}^h is defined essentially to be all forms (with appropriate symmetry) divided by these "hyperbolic" ones. Therefore if the equivalence class of the relative homology form is trivial in L_{2k}^h, it is equivalent to a hyperbolic form and the algebraic problem can be solved.

Supposing the algebraic obstruction is trivial leads to the embedding problem: given a hyperbolic summand of the middle dimensional homology, with basis represented by immersed spheres, find half a basis represented by disjoint embedded spheres. In higher dimensions the intersection and selfintersection conditions allow the use of Whitney moves to get framed embedded spheres. In dimension 4 the main embedding theorem (at least up to s-cobordism) is required.

11.3B Explanations: $L_*(\pi_1 X, \omega)$ are groups of forms or automorphisms of forms (Wall [**1**]), with variations L^h or L^s depending on the structure set S^h or S^s being studied. These groups are rather complicated, and have been the object of much study.

"Exactness" at $S_{\mathrm{TOP}}(X, N)$ means the group L_{n+1} acts on the structure set, and the orbits of this action are exactly point inverses of the function $S_{\mathrm{TOP}}(X, N) \longrightarrow NM_{\mathrm{TOP}}(X, N)$. NM_{TOP} denotes the set of normal maps, and exactness here means that the image of $S_{\mathrm{TOP}}(X, N)$ is exactly the preimage of $0 \in L_n(\pi_1 X)$.

Suppose (X, N) is a Poincaré pair. A *normal map* is a map from a manifold $(M, \partial M) \longrightarrow (X, N)$ together with a topological bundle over X, and an isomorphism of the pullback of this bundle to M with the stable normal bundle of M. The normal map is "relative" if N is a manifold, the map $\partial M \longrightarrow N$ is a homeomorphism, and the bundle map over ∂M is induced from this. If N is a manifold we define $NM(X, N)$ to be the set of normal bordism classes of degree 1 relative normal maps. This can also be described as the set of topological bundle structures on the normal spherical fibration of X, extending the given structure on N (Browder [**1**, §2]). These correspond to homotopy classes of lifts of the classifying map $X \longrightarrow B_{\mathrm{G}}$ into B_{TOP}. The fiber of this is G/TOP, which has low homotopy groups $\pi_2 = \mathbf{Z}/2$, $\pi_4 = \mathbf{Z}$, and trivial first k-invariant.

From this we conclude: if X is 4-dimensional the obstruction to nonemptyness of $NM(X, N)$ is an element of $H^3(X, N; \mathbf{Z}/2)$. If the set is nonempty then choosing an element as basepoint gives an identification with a set of homotopy classes $NM(X, N) \simeq [X, N; \mathrm{G/TOP}, *]$. This is $H^2(X, N; \mathbf{Z}/2) \oplus H^4(X, N; \mathbf{Z})$ when X is 4-dimensional.

If X is orientable ($\omega = 0$) then the sequence can be simplified somewhat. There is an action of the group $\mathbf{Z}$ on both the normal map set and the surgery group, and the surgery obstruction is equivariant with respect to this action. On $NM_{\mathrm{TOP}}(X, N)$ the action is given by connected sum with a standard normal map $\|E_8\| \longrightarrow S^4$, whose surgery obstruction is a generator in $L_4(\{1\})$. Dividing by this action gives the *reduced surgery sequence*. In the surgery group the homomorphisms $1 \longrightarrow \pi_1 X \longrightarrow 1$ induce a splitting $L_4(\pi_1 X, 0) = \tilde{L}_4(\pi_1 X, 0) \oplus L_4(\{1\})$.

The action of $\mathbf{Z}$ is given by the standard group structure and the inclusion of $\mathbf{Z} \simeq L_4(\{1\})$. In the normal map set dividing by $\mathbf{Z}$ kills the $H^4(X, N; \mathbf{Z})$ summand. Therefore there is an "exact" sequence

$$L_5^h(\pi_1 X, 0) \to S_{\mathrm{TOP}}(X, N) \to H^2(X, N; \mathbf{Z}/2) \to \tilde{L}_4(\pi_1 X, 0)$$

for an orientable 4-dimensional Poincaré complex with a distinguished topological bundle structure on the normal fibration, and good fundamental group.

Finally we note that if X is a manifold then the exact sequence extends to the left by the exact sequence for $X \times I$;

$$\cdots \to S_{\mathrm{TOP}}^h(X \times I, \partial(X \times I)) \to NM_{\mathrm{TOP}}(X \times I)) \to L_{n+1}^h(\pi_1 X, \omega) \to$$

$$\to S_{\mathrm{TOP}}^h(X, N) \to NM_{\mathrm{TOP}}(X, N) \to \cdots$$

The sets for $X \times I$ have natural group structures obtained by glueing in the I coordinate. With respect to these the left part of the sequence is exact as a sequence of groups.

11.3C The Surgery Program: Exactness of the surgery sequence suggests an approach to the classification of manifolds, namely

(1) Classify Poincaré complexes,
(2) determine which of these have a topological reduction of the normal fibration, and which reductions have trivial surgery obstruction in $L_n(\pi_1 X)$, and
(3) determine the action of L_{n+1} on $S_{\mathrm{TOP}}(X)$.

Actually this program leads to $S_{\mathrm{TOP}}(X)$, which describes manifolds with given maps to reference spaces X. If manifolds without reference maps are the objective, then it is also necessary to understand the effect of homotopy self-equivalences of X on $S_{\mathrm{TOP}}(X)$.

Closed simply-connected 4-manifolds were first classified using this program (Freedman [**2**, Theorem 1.5]), and we will discuss this further in the next section. Simply-connected 4-manifolds with boundary have been classified this way by Vogel [**1**], and certain restricted classes of nonsimply connected 4-manifolds with finite fundamental group have been studied by Hambleton and Kreck [**1**]. The program works well for some classes of aspherical manifolds (see 11.5). We carry out most of the program for 4-manifolds homotopy equivalent to 1-complexes in 11.6. Generally the homotopy theory and algebraic problems are quite complicated, and the program is difficult to implement.

11.3D Relative surgery. There is a relative version of surgery in which part of the boundary is allowed to vary (see Wall [**1**, §3]). This also extends to low dimensions, as long as the 4-dimensional ones have good fundamental groups. The most basic case is the "$\pi - \pi$ theorem" in which the boundary fundamental group is isomorphic to that of the interior. In this case there are no obstructions (Wall [**1**, §4]). A low-dimensional version of this is given in section 11.9B.

11.4 Simply connected 4-manifolds

Suppose (X, N) is a simply-connected 4-dimensional Poincaré pair, with N a 3-manifold. There is a topological reduction for the normal fibration because the obstruction lies in the trivial group; $H^3(X, M; \mathbf{Z}/2)$ is dual to $H_1(X; \mathbf{Z}/2) = 0$. The surgery groups are $L_5(\{1\}) \simeq \tilde{L}_4(\{1\}) = 0$, so the reduced surgery sequence gives a bijection

$$S_{\mathrm{TOP}}(X, N) \simeq H^2(X, N; \mathbf{Z}/2).$$

The conclusion is that simply-connected manifolds can be classified by classifying simply-connected Poincaré spaces with 3-manifold boundary.

If the boundary is empty, Poincaré spaces are characterized by the form on the middle homology. Therefore if (H, λ) is a form there is X with this form and $S_{\mathrm{TOP}}(X, N) \simeq H \otimes (\mathbf{Z}/2)$.

This calculation determines the set of homotopy equivalences $M \rightarrow X$, with two identified if there is a homeomorphism between the domain manifolds which homotopy commutes with the maps to X. Note this is a bit different than the conclusion of the classification theorem 10.1, which asserts that there are at most two manifolds with a given intersection form. The equivalence relation used there is homeomorphism, with no homotopy commutativity requirement. The classification theorem was first proved by starting with this calculation of $S_{\mathrm{TOP}}(X)$, and then constructing self-equivalences of X to realize at least half of the invariants in $H^2(X; \mathbf{Z}/2)$. The construction of self-equivalences was first indicated by Wall [**1**, p. 237]; see Cochran and Habegger [**1**] for a corrected and expanded version.

Simply-connected 4-manifolds with (fixed) nonempty boundary have been classified by Vogel [**1**], using the same techniques.

Contractible manifolds, which we have obtained from plus constructions, can also be found with the surgery sequence. Suppose N is a homology sphere, then $(\mathrm{cone}N, N)$ is a Poincaré pair. Since the cohomology $H^2(\mathrm{cone}N, N; \mathbf{Z}/2)$ is trivial, the bijection established above shows that $S_{\mathrm{TOP}}(\mathrm{cone}N, N; \mathbf{Z}/2)$ is a single point. Therefore there is a contractible manifold with boundary N, and this manifold is unique up to homeomorphism.

11.5 Aspherical 4-manifolds

A manifold is aspherical if $\pi_i M = 0$ for $i > 1$. Homotopically the only invariant of such a manifold is its fundamental group. Often the same is true topologically;

Theorem. *Suppose $f\colon M \to N$ is a homotopy equivalence of compact aspherical 4-manifolds with poly-(finite or cyclic) fundamental groups, which restricts to a homeomorphism of boundaries. Then f is homotopic rel ∂M to a homeomorphism.*

This is proved for higher dimensions by Farrell and Jones [1], extending earlier work of Farrell and Hsiang [1]. They show that the surgery obstruction $NM_{\mathrm{TOP}}(N, \partial N) \to L_n(\mathbf{Z}\pi_1 N, \omega)$ is an isomorphism, so the surgery exact sequence shows $S_{\mathrm{TOP}}(N, \partial N)$ consists of one element. Since the exact sequence is valid in dimension 4 when fundamental groups are good, and poly-(finite or cyclic) groups are good, this argument extends to dimension 4. ∎

Tori are particularly useful cases: T^4, $T^3 \times I$, $T^2 \times D^2$, and $S^1 \times D^3$ are all topologically unique (rel boundary).

We note that Farrell and Jones [2] have proved the analogous uniqueness result for hyperbolic manifolds, another aspherical class, in dimensions ≥ 5. However hyperbolic groups are not known to be good, so we can draw no conclusions about uniqueness of hyperbolic 4-manifolds. There is a similar problem with Poincaré spaces with the homotopy type of a 1-complex, described in the next section.

11.6 Homotopy 1-complexes

In this section we take as far as possible the surgery program for Poincaré pairs (X, Y) with X homotopy equivalent to a 1-complex; the complexes are characterized and the obstructions described. These have free fundamental groups, which are "good" only when rank 1, so we only obtain manifold results for complexes homotopic to S^1. These are applied to knots in the next section. The rank greater than one cases will be used in Chapter 12 to give reformulations of the embedding problems in terms of links, and Poincaré transversality.

11.6A Proposition. *Suppose (X, N) is Poincaré, N is an orientable 3-manifold, and X is equivalent to a 1-complex. Then manifolds with boundary N and $\mathbf{Z}[\pi_1 X]$-homology equivalent to X are unique up to homology s-cobordism rel boundary. If $\pi_1 X$ is good then there is a manifold homotopy equivalent to X, and it is unique up to homeomorphism rel boundary.*

The s-cobordism statement and the plus construction implies that the structure set $S_{\text{TOP}}(X, N)$ consists of at most one point, with either the homology or homotopy definitions. As mentioned above, $\pi_1 X$ is only known to be good if X is contractible or equivalent to S^1. In these cases the structure set will be shown to be nonempty, so is exactly one point. Existence in the other cases will be analysed further in 12.3.

Proof: The Whitehead group of a free group is trivial, so there is no distinction between L^h and L^s, h- and s-cobordisms, etc.

Suppose M and M' are manifolds with boundary N and $\mathbf{Z}[\pi_1 X]$ equivalent to X. The reduced normal map set is $H^3(X, N; \mathbf{Z}/2) = 0$ so M and M' are normally bordant. Denote by Y the union of mapping cylinders of the maps to X, then the normal bordism defines a normal map to $(Y, M \cup M' \cup N \times I)$. This is a 5-dimensional surgery problem, with obstruction in $L_5^s(\pi_1 X)$. We claim it can be modified to give a surgery problem with trivial obstruction.

If π is a free group of rank k then there is an isomorphism $(L_4(1))^k \to L_5(\pi)$ given by taking a 4-dimensional problem in the ith factor, multiplying by S^1, and mapping to the ith generator in π. In particular generators are obtained by $S^1 \times (\|E_8\| \to S^4)$. Consequently there is an action of $L_5(\pi)$ on the surgery obstruction function $NM_{\text{TOP}}(Y, \partial Y) \to L_5(\pi)$, defined on generators by connected sum along loops. As with the reduced surgery sequence in the 1-connected case (11.4) this implies normal maps can be modified on the interior to have trivial obstruction.

Applying the (5-dimensional) surgery theorem to a normal map with trivial obstruction produces an s-cobordism from M to M'. This proves uniqueness up to s-cobordism. Now consider the 4-dimensional cases of the theorem.

We will show in Lemma 11.6B that there is always a normal map to (X, N). $H^2(X, N; \mathbf{Z}/2) \simeq H_2(X; \mathbf{Z}/2) = 0$, and the reduced surgery group $\tilde{L}_4(\pi_1 X)$ is also trivial. The reduced surgery sequence therefore implies that there is a unique normal map (up to bordism) with trivial surgery obstruction. If $\pi_1 X$ is good, so the surgery sequence is exact, this implies the structure set $S_{\text{TOP}}(X, N)$ is nonempty; there is a manifold equivalent to X with boundary N.

We have already seen that manifolds equivalent to X are unique up to s-cobordism. If $\pi_1 X$ is good the s-cobordism theorem is valid, so s-cobordant manifolds are homeomorphic. ∎

We tie up the loose end left hanging above:

11.6B Lemma. *Suppose (X, N) is Poincaré, X is equivalent to a 1-complex, and N is an orientable 3-manifold. Then there is a (smooth) normal map $(M, N) \to (X, N)$.*

Proof: Thinking of X as coming from a link complement (as in 11.7) the proof corresponds to identifying the normal map obstruction with the Arf invariants of the component knots. These invariants vanish because the knots are algebraically slice.

Every orientable 3-manifold has a framing with respect to which it is a framed boundary (since the J-homomorphism $\pi_3(SO) \to \pi_3^s$ is onto, see Kirby [2]). Let θ be such a framing for N, and consider the bordism class $[N, \theta] \in \tilde{\Omega}_3^{fr}(X) = H_1(X; \mathbf{Z}/2)$. The lemma is proved by showing this is trivial; a framed bounding manifold $(M, N) \to (X, N)$ gives the required normal map.

Let $X \to \vee^k S^1$ be a homotopy equivalence. Then there is an isomorphism $\Omega_3^{fr}(X) \to \Omega_3^{fr}(*) \oplus \left(\Omega_2^{fr}(*)\right)^k$ given by the point map in the first factor, and bordism classes of transverse inverse images of points in the circles in the other factors. The image of $[N, \theta]$ in the first factor is trivial by choice of θ, so it is sufficient to show the inverse images are framed boundaries.

Let S_j denote the inverse image of a point in the jth circle, and arrange it to be connected. Let $\hat{N}$ denote the covering space of N pulled back from the universal cover of $\vee^k S^1$. S_j lifts to $\hat{N}$ and splits it into two pieces, say $\hat{N}_+ \cup \hat{N}_-$. The perfect kernel hypothesis implies that $H_1(\hat{N}) = 0$, and dually $H_2(\hat{N}) = 0$, so the Mayer-Vietoris sequence for the splitting gives an isomorphism $H_1(S_j) \simeq H_1(\hat{N}_+) \oplus H_1(\hat{N}_-)$.

The intersection form is trivial on the kernels of $H_1(S_j) \to H_1(\hat{N}_\pm)$ so each of these kernels is half the rank of $H_1(S_j)$ and bases can be represented by disjoint embedded loops in S_j. Choose embedded loops representing a basis for the kernel of $H_1(S_j) \to H_1(\hat{N}_+)$. These bound orientable surfaces mapping to $\hat{N}_+$. Since the framing of S_j extends over $\hat{N}_+$, this implies the framings of the boundary loops is compatible with the framing of S_j. This means we can construct a framed bordism of S_j by attaching 2-handles to $S_j \times I$ on these framed loops in $S_j \times \{1\}$. Since the loops are half a basis for $H_1(S_j)$, this gives a bordism to S^2. Since $\pi_2(SO) = 0$ the framing extends over the bordism obtained by adding D^3 to this S^2. The result is a framed null bordism of S_j, as required. ∎

The last topic of the section is a structure theorem for such Poincaré pairs.

11.6C Proposition. *Suppose Y is a connected orientable 3-dimensional Poincaré space.*

(1) **Recognition:** *If (X, Y) is Poincaré, then X is a homotopy 1-complex if and only if $\pi_1 X$ is free, $\pi_1(X, Y) = 0$, and $H_i(X; \mathbf{Z}) = 0$ for $i > 1$.*

(2) **Existence:** *There is a Poincaré pair (X, Y) with X a homotopy 1-complex if and only if there is a homomorphism with perfect kernel of $\pi_1 Y$ onto a free group, and then $\pi_1 Y \to \pi_1 X$ is such a homomorphism.*

(3) **Uniqueness:** *Poincaré pairs (X,Y) with X a homotopy 1-complex are determined up to homotopy by Y.*

Proof: Suppose (X, Y) is Poincaré, as in (1). Let π denote $\pi_1 X$, and let $f \colon \vee^k S^1 \to X$ induce an isomorphism on π_1. We show f is a homotopy equivalence by showing that $H_i(f; \mathbf{Z}[\pi]) = 0$ for all i, or equivalently $H_i(X; \mathbf{Z}[\pi]) = 0$ if $i > 0$.

The first step is to show that $H_i(X; \mathbf{Z}[\pi]) = 0$ if $i > 2$, and $H^3(X; A) = 0$ for any $\mathbf{Z}\pi$-module A. $H_i(X; \mathbf{Z}[\pi])$ is dual to $H^{4-i}(X, Y; \mathbf{Z}[\pi])$, which vanishes for $i > 4$ for dimension reasons, and for $i = 4$ by connectivity. $H^3(X; A)$ and $H_3(X; \mathbf{Z}[\pi])$ are dual to $H_1(X, Y; A)$ and $H^1(X, Y; \mathbf{Z}[\pi])$ respectively. Since $\pi_1(X, Y)$ is trivial the chain complex $C_*(X, Y; \mathbf{Z}\pi)$ is contractible up through dimension 1, which implies the 1-dimensional homology and cohomology with any coefficients vanishes. (It should also be clear that similar duality considerations show the conditions of (1) are necessary.)

This shows the only possibly nontrivial relative group of f is H_2. Let C_* denote the relative chain complex of f, with $\mathbf{Z}[\pi]$ coefficients. According to Wall [3] the vanishing of the other homology, and of $H^3(C_*; A)$ for any A, implies $H_2(C_*)$ is stably free as a $\mathbf{Z}[\pi]$ module. The tensor product $H_2(f; \mathbf{Z}[\pi]) \otimes_{\mathbf{Z}[\pi]} \mathbf{Z}$ is the integral homology $H_2(f; \mathbf{Z})$, which is assumed to vanish. For stably free modules this implies vanishing, so $H_2(f; \mathbf{Z}[\pi]) = 0$, and f is a homotopy equivalence.

We extract the relevant part of Wall's argument. The boundary homomorphism $d_3 \colon C_3 \to \operatorname{im} d_3$ defines a homology class in $H^3(C_*; \operatorname{im} d_3)$, which has been shown to vanish. Therefore d_3 is a coboundary; there is a homomorphism $\theta \colon C_2 \to \operatorname{im} d_3$ so that $\theta d_3 = d_3$. This implies C_2 splits as $\operatorname{im} d_3 \oplus B$, and C_* has the same homology as the projective complex

$$0 \to B \xrightarrow{d_2} C_1 \xrightarrow{d_1} C_0.$$

H_0 and H_1 also vanish, which implies the image of d_2 is projective. This implies $B = H_2 \oplus \operatorname{im} d_2$, and therefore H_2 is projective. A little more care shows additionally that it is stably free, but in fact projective is sufficient in the argument above. Indeed, for π a free group finitely generated projective $\mathbf{Z}\pi$ modules are free (P. M. Cohen [1]).

Now consider (2). If (X, Y) is Poincaré then we saw above that $\pi_1 Y \to \pi_1 X$ is onto. $\pi_1 X \ (= \pi)$ is free, so it is sufficient to see the kernel of this is perfect. The abelianization of the kernel is $H_1(Y; \mathbf{Z}[\pi]) = H_2(X, Y; \mathbf{Z}[\pi]) \simeq H^2(X; \mathbf{Z}[\pi])$. But since X is a homotopy 1-complex H^2 vanishes, so the kernel is perfect.

For the converse, suppose $\pi_1 Y \to \pi$ is a homomorphism onto a free group. Realize this by a map $Y \to \vee^k S^1$, and let X denote the mapping cylinder. Choose as a fundamental class $[X, Y] \in H_4(X, Y; \mathbf{Z})$ whose boundary is $[Y]$, then we want to show

$$\cap[X, Y] : H^i(X; \mathbf{Z}[\pi]) \to H_{4-i}(X, Y; \mathbf{Z}[\pi])$$

is an isomorphism. This is the case when $i = 0$ since it is taken by the boundary to the isomorphism $H^0(Y) \simeq H_3(Y)$. For other values of i the group on the left vanishes, so the object is to show that $H_j(X, Y; \mathbf{Z}[\pi]) = 0$ when $j < 4$. When $j = 0, 1$, this is the case because π_1 is onto. For $j = 2, 3$, the boundary homomorphism to $H_{j-1}(Y; \mathbf{Z}[\pi])$ is an isomorphism. $H_1(Y; \mathbf{Z}[\pi])$ is the abelianization of the kernel of $\pi_1 Y \to \pi_1 X$, so is trivial. This also implies $H^1(Y; \mathbf{Z}[\pi]) = 0$. This cohomology is dual to the other homology group required to vanish, so the proof is complete.

Finally we prove the uniqueness statement (3). A map to a 1-complex is determined up to homotopy by the homomorphism on π_1, so the objective is to show that if X, X' are two such Poincaré spaces that there is an isomorphism $\pi_1 X \to \pi_1 X'$ which commutes with the homomorphism from $\pi_1 Y$. Since $\pi_1 Y$ is onto, this is equivalent to the kernels being equal. We claim that in each case the kernel is the intersection of the lower central series of $\pi_1 Y$.

Any perfect subgroup lies in the intersection of the lower central series, so the kernel is contained in the intersection. On the other hand the intersection of the lower central series of a free group is trivial, and homomorphisms preserve these intersections, so the kernel must contain the intersection. This shows they are equal, and completes the proof. ∎

11.7 Knots and links

Surgery can be used to study knots. In higher dimensions this is done with the homology surgery of Cappell and Shaneson [1], which does not extend to dimension 4 (see 11.8 below). However we can use the material developed above to identify "trivial" knots in S^3 and S^4.

11.7A Theorem. *A locally flat embedding $f : S^2 \to S^4$ is unknotted (isotopic to the standard embedding) if and only if $\pi_1(S^4 - f(S^2)) \simeq \mathbf{Z}$.*

Proof: Since the embedding is locally flat it has a normal vector bundle (9.3). Such bundles over S^2 are determined by the Euler number, which is the algebraic intersection between the 0-section and any other section transverse to it. Since intersection numbers are defined on homology classes, and $H_2(S^4) = 0$, this number (therefore the bundle) is trivial. Thus there is a neighborhood homeomorphic to $S^2 \times D^2$. Let M denote the complement of the interior of this neighborhood. Alexander duality shows $H^i(M; \mathbf{Z}) = 0$ for $i > 1$, so the assumption that $\pi_1 M \simeq \mathbf{Z}$ implies, by 11.6C(1) that M is homotopy equivalent to S^1. According to 11.6A such manifolds are uniquely determined by their boundaries. $\partial M = S^2 \times S^1$ so $M \simeq D^3 \times S^1$. Fitting this together with $S^2 \times D^2$ gives a homeomorphism $S^4 = S^2 \times D^2 \cup M \rightarrow S^2 \times D^2 \cup D^3 \times S^1 = S^4$, which takes $f(S^2)$ to the standard S^2.

This homeomorphism is isotopic to the identity, proving the corollary. To get an isotopy arrange f to coincide with the standard embedding on a small disk, so the resulting $S^4 \rightarrow S^4$ is the identity on a ball. The Alexander isotopy can then be applied to the homeomorphism on the complementary ball. ∎

The part specific to dimension 4 is the application of 11.6C. The same proof shows for arbitrary $n \geq 2$ that an embedding $f \colon S^n \rightarrow S^{n+2}$ is unknotted if and only if the complement is homotopy equivalent to S^1. However in higher dimensions there are many knots whose complements have fundamental group $\mathbf{Z}$, but not the homotopy type of S^1.

We call an embedding of S^1 in a homology 3-sphere N "$\mathbf{Z}$-slice" if it extends to a (locally flat) embedding of D^2 in the contractible 4-manifold bounding N, so that the complement has fundamental group $\mathbf{Z}$. We note that a general slice complement has a homomorphism $\pi_1 \rightarrow \mathbf{Z}$ whose kernel is the commutator subgroup, since $H_1(\text{complement}) \simeq \mathbf{Z}$ by Alexander duality. If the kernel is perfect we can get a $\mathbf{Z}$-slice using the plus construction, but usually the kernel will not be perfect.

11.7B Theorem. *An embedding $f \colon S^1 \rightarrow N$, N a 3-dimensional manifold homology sphere, is $\mathbf{Z}$-slice if and only if the the natural homomorphism $\pi_1(N - f(S^1)) \rightarrow \mathbf{Z}$ has perfect kernel, or equivalently, the Alexander polynomial of the knot is 1.*

Proof: We frame the normal bundle of f by: represent the generator of $H^1(N - f(S^1); \mathbf{Z})$ by a map $N - f(S^1) \rightarrow S^1$. This is a homotopy equivalence on the linking circle of f, so determines a fiber homotopy trivialization of the normal circle bundle. This determines a framing, because the classifying space for homotopy S^1 fibrations is homotopy equivalent to the classifying space for 2-dimensional vector bundles. Note this also implies we can assume that near $f(S^1)$ the map $N - f(S^1) \rightarrow S^1$ comes from the framing.

This framing is related to a "Seifert surface" for the knot. Make the map $N - f(S^1) \rightarrow S^1$ transverse to a point, then the preimage is an orientable surface. This surface intersects spheres in the normal bundle in a section of the framing, so adding $f(S^1)$ makes it a compact surface with $f(S^1)$ as boundary. Conversely such a surface splits the normal bundle of $f(S^1)$ into a sum of orientable line bundles, which determines the framing.

Now suppose f is **Z**-slice, with slice $F \colon D^2 \to M$. Denote the kernel of $\pi_1(N - f(S^1)) \to \mathbf{Z}$ by K. The complement of an open regular neighborhood of F is a compact manifold, with $\pi_1 \simeq \mathbf{Z}$. Excision shows that the **Z** homology vanishes above dimension 1, so by 11.6C(1) it is homotopy equivalent to S^1. But then by 11.6C(2) the homomorphism from π_1 of the boundary to **Z** has perfect kernel. This boundary is obtained from the complement of an open regular neighborhood of f in N by attaching $D^2 \times S^1$, with the D^2 attached on the boundary of the Seifert surface. Therefore $K/(\alpha)$ is perfect, where (α) denotes the normal closure of the boundary of the Seifert surface. But this surface lifts into the cover corresponding to K, which implies α is a commutator in K. This implies K is perfect, as required for the theorem.

For the converse, suppose the kernel is perfect. Define N_1 to be obtained from the complement of an open regular neighborhood of f in N by attaching $D^2 \times S^1$, with the D^2 attached on the boundary of the Seifert surface (in the above this gives the boundary of the complement of the slice). The map extends to $N_1 \to S^1$, and since the kernel of this is obtained from K by adding a relation, it is also perfect. According to 11.6C(2) this implies that the mapping cylinder is a Poincaré pair. Then by 11.6A this is homotopy equivalent to a manifold, since **Z** is good. Denote this manifold by M.

Attach a 2-handle to M on $D^2 \times S^1 \subset N_1$. This gives a contractible manifold with boundary N. The knot f is the boundary of the dual of this 2-handle, so it is **Z**-slice.

The equivalence of perfectness of the kernel with the Alexander polynomial being 1 is a result of Crowell [1]. Basically, the Alexander polynomial is the determinant of a relation matrix for $H_1(M_1; \mathbf{Z}[\mathbf{Z}])$ (the abelianization of the kernel) as a module over $\mathbf{Z}[\mathbf{Z}]$. The determinant is 1 (after normalization) if and only if the relations generate the entire module. This means the kernel abelianizes to give the trivial group, so is perfect. ∎

This result can also be proved by applying the compactly supported plus construction to complements product with I, as in 11.1C. The compactly supported plus construction is given in the exercise after 11.1A. An equivariant version of 11.7B is contained in the exercise after 11.1C.

11.7C Links. The same arguments apply to links, except the fundamental groups are free (nonabelian) groups, which are not known to be "good."

Define a *good boundary link* to be a collection of locally flat circles embedded in a 3-manifold, with a homomorphism with perfect kernel of π_1 of the complement to a free group, so that the images of linking circles form a set of generators. The term "boundary link" comes from the fact that such a homomorphism to a free group (with no kernel condition) is equivalent to the existence of a collection of disjoint orientable surfaces with the link as boundary. A link is *free-slice* (extending "**Z**-slice") if it extends to disjoint embedded disks in a contractible 4-manifold, with free complementary π_1.

Corollary. *Suppose "surgery works" (the sequence is exact) for 4-dimensional Poincaré pairs (X, Y) with X a homotopy 1-complex. Then a link in a 3-manifold is free-slice if and only if it is a good boundary link. Suppose "the s-cobordism theorem works" (rel boundary) for 4-manifolds equivalent to 1-complexes. Then a link in S^4 is standard if and only if the complementary fundamental group is free.*

Proof: For the first part, the proof of 11.7B works after subsituting $\vee^k S^1$ for S^1. The second part similarly follows from the proof of 11.7A, with the observation that the construction of an s-cobordism from the link complement to the standard complement is obtained by 5-dimensional surgery, without any "goodness" problems. "Goodness" of the group was used to find a product structure on this s-cobordism, and therefore a homeomorphism between the ends. ∎

The surgery data for good boundary links comes from Freedman [3], which also gives singular slices. See section 12.3 for examples of good boundary links, and a converse to the corollary.

11.8 Homology equivalence

Suppose $\mathcal{F}\colon \mathbf{Z}[\pi_1 X] \to \Lambda$ is a ring homomorphism. Cappell and Shaneson [1] have developed a surgery theory to study $M \to X$ which are homology equivalences with Λ coefficients. The motivating example comes from knot theory: if $f\colon S^n \to S^{n+2}$ is an embedding then there is a map $(S^{n+2} - f(S^n)) \to S^1$ which is a **Z** homology equivalence, but (when $n > 4$) usually is not a $\mathbf{Z}[\mathbf{Z}]$ homology equivalence even if $\pi_1(S^{n+2} - f(S^n)) \simeq \mathbf{Z}$.

This theory does not extend to dimension 4, even for good fundamental groups. We describe a counterexample, and discuss the technical obstacles to the construction.

The counterexample comes from knot theory. If homology surgery for $\pi_1 \simeq \mathbf{Z}$ and coefficients $\mathbf{Z}[\mathbf{Z}] \to \mathbf{Z}$ were to work, then it could be applied to study slices for knots in S^3. The techniques used in higher dimensions would show that if a knot is "algebraically slice" (see Casson and Gordon [1]) then it is at least weakly slice in the sense that it bounds a disk embedded in a manifold with the homology of D^4. In higher dimensions one would perform a plus construction on the complement to make the manifold simply connected, therefore homeomorphic to the ball. Here it might not be possible to use the plus construction because the fundamental group of the complement may not be good. But in fact Casson and Gordon [1] show there are algebraically slice knots in S^3 which are not even weakly slice. Therefore the homology surgery theory cannot be valid in dimension 4.

We explain the difficulty in the proof of the homology surgery theorem, and pose the problem of finding a version which works. In the proof of theorem 1.7 of Cappell-Shaneson [1, p.293] they show vanishing of the obstruction gives a collection of immersed spheres $S^k \to M^{2k}$ with trivial (in $\mathbf{Z}[\pi_1 M]$) algebraic intersections and selfintersections. The theorem is proved by finding a regular homotopy to disjoint embeddings, using the Whitney procedure when $k > 2$. The embedding theorems in dimension 4 require algebraic dual classes, as well as algebraically trivial intersections. In ordinary ($\mathbf{Z}[\pi_1]$ coefficient) surgery the nonsingularity of the intersection form implies the existence of dual classes. In the homological situation the intersection form only becomes nonsingular over the ring Λ, so only Λ coefficient dual classes can be found. Only rarely can these be arranged to be $\mathbf{Z}[\pi_1]$ coefficient duals.

Homology surgery does work "stably" (after connected sums with $S^2 \times S^2$), because there is a stable embedding theorem which does not require duals.

11.9 Ends of 4-manifolds

A manifold has "ends" if it is not compact. In higher dimensions the principal result identified ends which have open collar neighborhoods. This result and the definitions are reviewed in 11.9A. The 4-dimensional analog, given in 11.9B, characterizes the existence of "weak collars." Weak collars themselves are characterized in 11.9C, and this result reformulates the classification problem for weakly collared ends in surgery terms. General tame ends (without weak collars) are shown to have periodic neighborhoods in 11.9F. This allows expression of the classification problem for these in terms of "proper surgery," in 11.9F. The surgery classification of ends is applied in the next section to the study of fixed points of group actions on 4-manifolds.

11.9A Background. Suppose X is a locally compact space, then a "neighborhood of the end" of X is a set whose complement has compact closure. We restrict to ends which are *connected*, in the sense that every neighborhood of the end contains a connected neighborhood. The results developed presented here extend in a formal manner to manifolds with finitely many ends. Further extensions, for example to "controlled" ends as in Quinn [2-4] will not be discussed.

The nicest ends are the ones with collars: neighborhoods of the form $N \times (0, 1]$ with N compact. The basic results involve comparison of collared ends with ones satisfying a homotopy condition.

The end of X is *tame* if there is a closed neighborhood U of the end, and a proper map $U \times (0, 1] \to X$ which is the inclusion on $U \times \{1\}$.

There is a lot packed into the properness requirement in this definition; each $U \times \{t\} \to X$ preserves ends, and given any neighborhood V of the end there is $t > 0$ such that the image of $U \times (0, t]$ is inside V. Note that an open collar is trivially tame: multiplication in the $(0, 1]$ coordinates provides the required map $N \times (0, 1] \times (0, 1] \to N \times (0, 1]$.

This definition is different from (and much simpler than) that used in the original treatment in Siebenmann's thesis. That definition involved maps pulling neighborhoods *away* from the end rather than toward it. This version is taken from Quinn [11, 2.1], where the older notion is called "reverse tameness." An end of a manifold is tame and has stable π_1 in the sense of Siebenmann, if and only if it is tame in the sense above and has neighborhoods with finitely presented π_1 (Quinn [11, prop. 2.14]).

Suppose M is a manifold with a tame connected end with finitely presented fundamental group π. Siebenmann associates to this an invariant $\sigma(\text{end}M) \in \tilde{K}_0(\mathbf{Z}\pi)$. Specifically, if U is a closed manifold neighborhood of the end which has a map as in the definition of tameness, then U is dominated by a finite complex and has a homomorphism $\pi_1(U) \to \pi$. Then $\sigma(\text{end}M)$ is the image of the Wall finiteness obstruction $\sigma(U) \in \tilde{K}_0(\mathbf{Z}\pi_1U)$, see Wall [3]. Seibenmann's result (extended to dimension 5 as in Quinn [4]) is:

End theorem. *Suppose M has a tame connected end with finitely presented fundamental group π. If $\dim M \geq 6$, or $\dim M = 5$ and π is good, then $\sigma(\text{end}M) \in \tilde{K}_0(\mathbf{Z}\pi)$ vanishes if and only if there is a collar neighborhood of the end.*

Partial extensions to dimension 4 and 3 are given in 11.9B. Before considering this we discuss the homotopy structure of tame ends.

Suppose X is locally compact, and denote by E the space of continuous proper maps $(0, 1] \to X$. The *homotopy collar* of the end is this space together with the evaluation map $e \colon E \times (0, 1] \to X$. (This is called the "homotopy completion" in Quinn [**3**, §7.8], and is the "homotopy link" of ∞ in the one-point compactification X^∞ in Quinn [**11**]).

It is simple to see that a closed neighborhood of the end has homotopy equivalent homotopy collar. Note the proper map $N \times (0,1] \to X$ in the definition of tameness defines a map into the homotopy collar $U \to E$. Up to homotopy this gives a factorization of the inclusion $U \subset X$ through e. Among other things this implies the fundamental group of the homotopy collar is the same as the "fundamental group of the end."

If an end has a geometric collar then the homotopy collar is homotopy equivalent to it. Since the complement of the interior of a geometric collar is a compact manifold, Poincaré duality is satisfied. The next proposition, from Quinn [**3**, §7.8], shows this is true even when there is no geometric collar.

Proposition. *Suppose M is a manifold with compact boundary and connected tame end, with finitely presented fundamental group. Let $e \colon E \to M$ denote the homotopy collar of the end, then the mapping cylinder $(M_e, \partial M \cup E)$ is a Poincaré space.*

It follows from this (or can be easily shown directly) that E is also dominated by finite complexes. We caution that the finiteness obstruction of the homotopy collar is not the same as the end obstruction, but the two are related. Let U be a closed manifold neighborhood of the end which satisfies the condition in the definition of tameness (so factors through E). The end obstruction is the finiteness obstruction of U, or equivalently the mapping cylinder U_e. If (X, Y) is a dominated Poincaré pair of dimension n then $\sigma Y = \sigma X + (-1)^{n-1}\bar{\omega}(\sigma X)$, in $\tilde{K}_0(\mathbf{Z}[\pi_1 X])$. Applying this to $(U_e, \partial U \cup E)$, and noting that ∂U makes no contribution since it is a manifold, we get $\sigma E = \sigma(\text{end}M) + (-1)^{n-1}\bar{\omega}(\sigma(\text{end}M))$.

11.9B Weak collars. Suppose M is a manifold with compact boundary. A *weak collar* for the end of M is a closed manifold neighborhood U of the end, which has a proper map $U \times (0,1] \to U$. The definition is like that of tameness, except that the map stays in U, not just in M.

The structure of weak collars will be investigated in the next section. Here we note only that the inclusion $\partial U \to U$ is a $\mathbf{Z}\pi_1 U$-homology equivalence, so these are homologically like collars. Also in higher dimensions existence of a weak collar implies existence of a genuine collar.

Weak end theorem. *Suppose M is a 4-manifold with compact boundary and a tame connected end with finitely presented good fundamental group. Then there is a weak collar of the end if and only if the obstruction $\sigma(\text{end}M) \in \tilde{K}_0(\mathbf{Z}\pi_1(\text{end}M))$ is trivial.*

We discuss the 3-dimensional situation before beginning the proof of the theorem (see also Hush and Price [1], Kalimizu [1]). A tame connected end of a 3-manifold has a weak collar. This follows from standard embedded surface theory and the result of Eckmann and Linnell [1] that a 2-dimensional Poincaré space has the homotopy type of a surface. The lack of an obstruction is consistent with the fact that $\tilde{K}_0(\mathbf{Z}\pi) = 0$ for π the fundamental group of a surface, but this fact is not used in the proof.

If U is a 3-dimensional weak collar then $\partial U \to U$ is a homotopy equivalence (because there are no nontrivial perfect subgroups of $\pi_1 \partial U$). In fact U is obtained from a genuine collar $\partial U \times (0,1]$ by connected sum at some discrete set with a collection of counterexamples to the 3-dimensional Poincaré conjecture. So if the Poincaré conjecture is true then the strong form of the end theorem is true for 3-manifolds.

Proof of the theorem: In higher dimensions the proof begins with a closed manifold neighborhood U of the end which maps to the homotopy collar. Then ∂U is modified inside M to make the inclusion $\partial U \to U$ a homotopy equivalence. In 4-manifolds there is not enough maneuvering room to use this approach. Instead we construct a compactly supported s-cobordism from M to a manifold containing a weak collar. The s-cobordism theorem implies the manifolds are homeomorphic, so M itself has a weak collar.

The construction is based on a version of the "$\pi - \pi$ lemma," which asserts there are no obstructions to surgery in certain relative situations. The basic statement is: suppose $(M; \partial_1 M, \partial_2 M) \to (X; \partial_1 X, \partial_2 X)$ is a degree 1 normal map, from a manifold to a finite Poincaré triad, such that $\partial_2 M \to \partial_2 X$ is a homotopy equivalence. Suppose also that $\pi_1 \partial_1 X \to \pi_1 X$ is an isomorphism. Then M is normally bordant rel $\partial_2 M$ to a homotopy equivalence of triads. (This is valid for $\dim X \geq 6$, or $= 5$ if $\pi_1 X$ is good.)

This basic statement will be modified in several ways. First since $\partial_2 M$ is held fixed it is sufficient for it to be a Poincaré space rather than a manifold. Specifically we work with an open manifold analogous to $M - \partial_2 M$ in the above statement, and suppose there is a map $\partial_2 X \to M$ so that the mapping cylinder $(M; \partial M, \partial_2 X)$ is a Poincaré triad. In the applications $\partial_2 X$ will be the homotopy collar of the end of M. The second modification involves enlarging the manifold to slightly weaken the "good" hypothesis on π_1. This is explained after the statement.

The $\pi - \pi$ lemma. Suppose $(X; \partial_1 X, \partial_2 X)$ is a Poincaré triad, M is a (noncompact) manifold, $(M, \partial M) \to (X, \partial_1 X)$ is a normal map and $e\colon (\partial_2 X, \partial \partial_2 X) \to (M, \partial M)$ is a homotopy lift of the inclusion in

$(X, \partial_1 X)$ so that the mapping cylinder $(M_e; \partial M_{\partial e}, \partial_2 X)$ is Poincaré. Suppose M, ∂M, X, and $\partial_i X$ are all homotopic to finite complexes and $\pi_1 \partial_1 X \to \pi_1 X$ is an isomorphism. Finally suppose that $\dim X \geq 4$, and if $\dim X = 4$, or 5, then either $\pi_1 X$ is good or there is an "enlargement" of M, X so the image of $\pi_1 X \to \pi_1 X'$ is good. Then there is a compactly supported normal bordism from M (or its enlargement) to $(M', \partial M')$ so that $M' \to X'$ is a homotopy equivalence.

The conclusion does not assert that $\partial M' \to \partial_1 X'$ is a homotopy equivalence, although it does follow from the other conditions that the map is a $\mathbf{Z}\pi_1 X'$ homology equivalence. Homotopy equivalence can be arranged via a plus construction if $\dim X \geq 5$.

A "compactly supported" bordism is one with a product structure outside a compact set. This is equivalent to having a handlebody structure with only finitely many handles.

An "enlargement" is obtained by glueing a manifold onto both M and X. To define this precisely we need a further subdivision of the boundary. Suppose $(X, \partial_1 X, \partial_2 X, \partial_3 X)$ is a Poincaré 4-ad, $(M; \partial_1 M, \partial_2 M) \to (X; \partial_1 X, \partial_2 X)$ is normal, and

$$e \colon (\partial_3 X; \partial_3 X \cap \partial_1 X, \partial_3 X \cap \partial_2 X) \to (M; \partial_1 M, \partial_2 M)$$

is a factorization of the inclusion which has Poincaré mapping cylinder. Suppose $\partial_2 M \to \partial_2 X$ is a homotopy equivalence, and there is a compact manifold V given with boundary $\partial V = \partial_1 V \cup \partial_2 M$. Then the "enlargement" of the data using V is obtained by adding V to both M and X; $(M \cup_{\partial_2 M} V - \partial_1 V, \partial_1 M) \to (X \cup_{\partial_2 X} V; \partial_1 X, \partial_3 X \cup \partial_1 V)$. The map $\partial_3 X \cup \partial_1 V \to M \cup (V - \partial_1 V)$ is given by e on $\partial_3 X$ and inclusion inside a collar on $\partial_1 V \to V - \partial_1 V$. The enlargement hypothesis in the lemma is that the image of $\pi_1 X \to \pi_1(X \cup V)$ is good. The conclusion is that there is a compactly supported normal bordism of $M \cup (V - \partial_1 V)$ to a manifold homotopy equivalent to $X \cup V$.

Proof of the $\pi - \pi$ lemma: This is based on the proof of the manifold version given in Wall [**1**, §4]. We indicate modifications required in this proof by low dimensions and the Poincaré hypotheses. The map $(M, \partial M) \to (X, \partial_1 X)$ will be denoted by f.

First suppose $\dim X = 4$, so we are in Wall's even dimensional case with $k = 2$. Using the hypothesis that M, X, etc. are equivalent to finite complexes, surgery can be done on M, ∂M to make ∂f 1-connected and f 2-connected. M now has the right π_1, but ∂M may not.

The hypothesis that the mapping cylinder of the factorization e is Poincaré can be used to define kernel groups $K_*(M, \partial M)$, etc. as in

Wall. Further Wall's argument (using the finiteness of M, X, etc.) shows that $K_2(M, \partial M)$ is stably free. Stabilize and choose a basis $\{e_i\}$. Since $\pi_1 M \longrightarrow \pi_1 X$ is an isomorphism this basis can be represented by maps $(D^2, S^1) \longrightarrow (M, \partial M)$ nullhomologous in $X, \partial_1 X$. The nullhomology together with the isomorphism $\pi_1 \partial_1 X \longrightarrow \pi_1 X$ implies the maps are nullhomotopic in $(X, \partial_1 X)$. As in Wall this specifies regular homotopy classes of framed immersions $f_i \colon (D^2, S^1) \longrightarrow (M, \partial M)$, homotopic to the $\{e_i\}$.

It is sufficient to show the $\{f_i\}$ are regularly homotopic to a π_1-null family of disjoint framed embeddings. The same homology calculations used by Wall but in the Poincaré triad $(M_e; \partial M, \partial_2 X)$ shows that the complement of these handles gives a manifold satisfying the conclusions of the lemma.

Since $\pi_1 \partial M \longrightarrow \pi_1 M$ is onto we can push boundaries of the f_i through each other, as in Wall [1, p. 53], to arbitrarily change intersection numbers among the f_i. Arrange in this way that the intersection numbers all vanish. We show that these immersions have framed immersed homologically dual spheres, so the embedding theorem 5.1B can be applied.

Since $K_2(M, \partial M)$ is free it is isomorphic to the dual $(K^2(M, \partial M))^*$, which is Poincaré dual to $K_2(M_e, \partial_2 X) \simeq K_2(M)$. Since $\pi_1 M = \pi_1 X$, $H_2(M; \mathbf{Z}\pi_1 X) = \pi_2(M)$ and so the basis for $K_2(M) \subset H_2(M; \mathbf{Z}\pi_1 X)$ is represented by maps $S^2 \longrightarrow M$. Also these maps are nullhomotopic in X, so since $M \longrightarrow X$ is normal these maps are represented by framed immersions. The isomorphism $K_2 M \simeq (K^2(M, \partial M))^* \simeq K_2(M, \partial M)$ is adjoint to the intersection form $\lambda \colon K_2(M) \otimes K_2(M, \partial M) \longrightarrow \mathbf{Z}\pi$, so these framed immersed spheres are algebraically dual to the immersions $\{f_i\}$.

If $\pi_1 X$ is good, then theorem 5.1B implies that the $\{f_i\}$ are regularly homotopic to π_1-null embeddings. As explained above these embeddings yield the desired manifold. If $\pi_1 X$ is not good, but there is an enlargement in which the image is good, then the embedding theorem gives embeddings in the larger manifold (see the exercise after 5.1A). Since enlargement involves adding homotopy equivalences the kernel $K_2(M, \partial M)$ does not change. Therefore the framed immersions in the enlarged manifold yield a modification of it which satisfies the conclusion of the lemma.

Now suppose $\dim X = 5$. This is the odd dimensional case of Wall [1, p. 41] with $k = 2$. Both $M \longrightarrow X$ and $\partial M \longrightarrow \partial_1 X$ can be made 2-connected, and $K_3(M, \partial_1 M)$ can be made free and based. Let $f_i \colon (D^3, S^2) \longrightarrow (M, \partial M)$ denote immersions representing a basis. Wall shows that if the boundaries $\partial f_i \colon S^2 \longrightarrow \partial M$ are regularly homotopic to disjoint framed embeddings, then the construction can be completed. (As above, Wall's homology calculations can be done using the hypothesis of duality in the triad $(M_e; \partial M_{\partial e}, \partial_2 X)$.)

As observed by Wall at the bottom of his page 41, the image of $\partial \colon K_3(M, \partial M) \to K_2(\partial M)$ is a subkernel. This means all intersection numbers among the $\{\partial f_i\}$ are trivial. Also there is a complementary subkernel, and in particular elements $A_j \in K_2(\partial M)$ so that $\lambda(\partial f_i, A_j) = 0$ if $i \neq j$, and $= 1$ if $i = j$. Since the A_j can be represented by framed immersed 2-spheres, if $\pi_1 M = \pi_1 X$ is good then corollary 5.1B implies that the $\{f_i\}$ are regularly homotopic to disjoint framed embeddings. As observed above this completes the argument.

If $\pi_1 X$ is not good, but has good image in an enlargement, then the corollary applies to give embeddings in the enlargement. As in the even-dimensional case we observe that passing to the enlargement does not change the K groups. Therefore the embeddings in the enlargement yield a normal bordism of the enlargement to a homotopy equivalence. ∎

We now return to the proof of the weak end theorem.

The objective is to construct a compactly supported s-cobordism of M to a manifold whose end has a weak collar. This will be done with two applications of the $\pi - \pi$ lemma; one 4-dimensional to construct the weak collar, and one 5-dimensional to construct the bordism.

Choose a closed connected manifold neighborhood of the end, V, which has a proper retraction to the end as in the definition of tameness. The inclusion $V \subset M$ factors up to homotopy through the homotopy collar $E \to M$. Let $U \subset V$ be another such neighborhood which retracts inside V. The inclusion $U \subset V$ also factors through E, so the image of $\pi_1 U \to \pi_1 V$ is the good group $\pi_1 E$. We will work primarily in U, reserving V to use as an enlargement.

Let $e \colon E \times (0, 1] \to U$ denote the homotopy collar, and let $U_e = U \cup_{E \times \{1\}} E \times [0, 1]$ denote the mapping cylinder of $e|E \times \{1\}$. Note $\mathrm{id} \cup e$ defines a map $U_e - E \times \{0\} \to U$. Since the end is tame, there is a neighborhood of the end whose inclusion into U factors through E up to homotopy. This and a proper map $U \to (0, 1]$ can be used to construct a homotopy from the inclusion $U \subset U_e$ rel ∂U to $f \colon U \to U_e$ with image in $U_e - E \times \{0\}$, so that composition with projection to $(0, 1]$ is proper. Roughly, f is a "proper over $(0, 1]$" homotopy inverse for $\mathrm{id} \cup e$.

Assume f is transverse to $E \times \{\frac{1}{2}\}$, then $f^{-1}((0, \frac{1}{2}]$ is a closed manifold neighborhood of the end. We set up the data for application of the $\pi - \pi$ lemma to $(f^{-1}(0, \frac{1}{2}], f^{-1}(\frac{1}{2}) \to (E \times [0, \frac{1}{2}]; E \times \{\frac{1}{2}\}, E \times \{0\})$.

This map is a normal map since the inclusion in M factors through E up to homotopy. Restricting the evaluation map gives a factorization $\hat{e} \colon E \to f^{-1}(0, \frac{1}{2}]$ homotopic to the homotopy collar of the end of

$f^{-1}(0, \frac{1}{2}]$. According to the proposition in 11.9A the mapping cylinder $(f^{-1}(0, \frac{1}{2}]_{\hat{e}}; \partial f^{-1}(0, \frac{1}{2}], E)$ is Poincaré, so $\hat{e}$ is the lift required for the lemma.

These spaces are all homotopy equivalent to finite complexes: the finiteness obstruction of $f^{-1}(0, \frac{1}{2}]$ is the end obstruction, which is assumed to vanish. According to the discussion at the end of 11.9A, the finiteness obstruction of E is obtained by symmetrization from this, so also vanishes.

Finally $\pi_1(E \times \{\frac{1}{2}\}) \rightarrow \pi_1(E \times [0, 1])$ is an isomorphism, and has been assumed to be good. Therefore the lemma implies there is a compactly supported normal bordism W, from $f^{-1}(0, \frac{1}{2}]$ to a manifold homotopy equivalent to E.

We fix some notation for the normal bordism W. There is a division of the boundary into $\partial_0 W \cup \partial_1 W \cup \partial_2 W$ and a map $g: W \rightarrow (E \times (0, \frac{1}{2}]) \times [0, 1]$ which takes $\partial_j W$ to $(E \times (0, \frac{1}{2}]) \times \{j\}$ for $j = 0, 1$, and $\partial_2 W$ to $(E \times \{\frac{1}{2}\}) \times [0, 1]$. Further the composition of this with the projection to $(0, \frac{1}{2}]$ is proper. $\partial_0 W = f^{-1}(0, \frac{1}{2}]$, the map $\partial_1 W \rightarrow E \times (0, \frac{1}{2}]$ is a homotopy equivalence, and outside some compact set W is a product $\partial_j W \times I$, where $j = 0$ or 1. Note that $\partial_1 W$ is a weak collar.

Next the data for a 5-dimensional application of the lemma will be set up. Define a Poincaré triad $(X; \partial_1 X, \partial_2 X)$ by $X = U_e \times I$, $\partial_1 X = U \cup_{E \times \{1\}} E \times [\frac{1}{2}, 1] \times \{1\}$, and $\partial_2 X = \partial U \times I \cup U_e \times \{0\} \cup (E \times \{0\}) \times I \cup (E \times [\frac{1}{2}, 1]) \times \{1\}$. These are all equivalent to finite complexes.

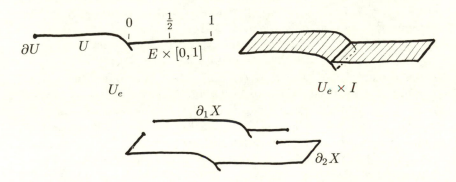

Let N be the union $U \times I \cup W$, where $\partial_0 W = f^{-1}(0, \frac{1}{2}]$ is identified with $f^{-1}(0, \frac{1}{2}] \times \{1\}$. Let $\partial_1 N = (U - f^{-1}(0, \frac{1}{2})) \times \{1\} \cup \partial_2 W$, and $\partial_2 N$ the rest of ∂N.

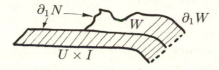

Then f and g define a map $(N; \partial_1 N, \partial_2 N) \to (X, \partial_1 X, \partial_2 X)$. Near the end of N it is a product $(\mathrm{end} M) \times I$, so the lifting $E \to U$ used above gives $E \times \{0\} \times I \to N$ so that the mapping cylinder in Poincaré. To get the situation of the $\pi - \pi$ lemma delete $\partial_2 N$, and restrict to $(N - \partial_2 N, \partial_1 N) \to (X, \partial_1 X)$.

Note $\pi_1 \partial_1 X$ and $\pi_1 X$ are both $\pi_1 U$, so all hypotheses of the lemma are satisfied except for the "goodness" of $\pi_1 X$. To fix this we enlarge. Recall there is a neighborhood of the end $V \supset U$ so that the image $\pi_1 U \to \pi_1 V$ is the good group $\pi_1(\mathrm{end} M)$. Enlarge both N and X by adding $(V - \mathrm{int} U) \times I$ on $\partial U \times I$ in $\partial_2 N$ and $\partial_2 X$. Since $N \to X$ is a homeomorphism on this part of the boundary, this gives an enlargement in the sense explained after the lemma. We still denote these enlargements by N, X, to avoid more notation.

The lemma produces a compactly supported normal bordism of $N - \partial_2 N$ to a manifold Z homotopy equivalent to X. Here we are not interested in the bordism itself, only Z. Adding $\partial_2 N$ back to Z (using the compactly supported hypothesis) gives a manifold $\hat{N}$. This is a compactly supported h-cobordism rel ∂U from $U \times \{0\}$ to $\partial \bar{Z} \cup \partial_1 W$. Note that $\partial \bar{Z} \cup \partial_1 W$ contains the weak collar $\partial_1 W$ of its end, so the theorem will be complete if we show this h-cobordism is a product.

At this point an adjustment is required to make the h-cobordism an s-cobordism. Since $\pi_1 \partial Z \to \pi_1 U$ is an isomorphism, we can attach to ∂Z an h-cobordism with appropriate torsion to cancel the torsion of $(\hat{N}, U \times \{0\})$. Continue to denote this by $\hat{N}$.

Finally we apply the s-cobordism theorem. $\hat{N}$ is a compactly supported s-cobordism (rel boundary), but its fundamental group is that of U, which may not be good. However $U \subset M$ factors through the homotopy collar which does have good fundamental group. Therefore extending $\hat{N}$ by the product cobordism on $M - U$ gives an s-cobordism of M with proper support, whose image in $\pi_1 M$ is good. The h-cobordism theorem 7.1C applies to this to give a product structure. In particular M is homeomorphic to a manifold with a weak collar, proving the theorem. ∎

Exercise Extend the weak end theorem to include boundary: suppose M is a 4-manifold with a tame connected end with good fundamental group. Suppose ∂M has a relative weak collar for its end (rather than being compact). Then show M has a weak collar if and only if the end obstruction vanishes. Here a "relative" weak collar is defined as before, and $U \cap \partial M$ is also required to be a weak collar. $\square$

11.9C Classification of weak collars. By the "data" of a weak collar U we will mean the compact manifold ∂U together with the homomorphism $\pi_1 \partial U \to \pi_1 U$. The substance of the next result is that this data characterizes weak collars.

Theorem. *Suppose π is a finitely presented good group.*

(1) *Suppose N is a compact 3-manifold. Then $(N, \pi_1 N \to \pi)$ is the data for a weak collar if and only if the homomorphism is onto and has perfect kernel.*

(2) *Any isomorphism of data associated to weak collars with fundamental group π is induced by a homeomorphism.*

(3) *Any $\mathbf{Z}\pi$-homology h-cobordism of data associated to weak collars with fundamental group π is (up to homology h-cobordism rel boundary) induced by a homeomorphism of neighborhoods of ends.*

In (2) a homeomorphism $U \simeq V$ gives a homeomorphism $\partial U \simeq \partial V$ together with a commutative diagram

$$
\begin{array}{ccc}
\pi_1 \partial U & \xrightarrow{\;\sim\;} & \pi_l \partial V \\
\downarrow & & \downarrow \\
\pi_1 U & \xrightarrow{\;\sim\;} & \pi_1 V,
\end{array}
$$

which is the "isomorphism of data" associated to the homeomorphism.

It will follow from the construction in (1) (or 11.9E below) that any neighborhood of the end of a weak collar U, contains a copy of U. Therefore homeomorphism of neighborhoods of ends of U and V, as in (3), gives an embedding $U \subset V$ whose complement has compact closure. This closure is a $\mathbf{Z}\pi$-homology h-cobordism from ∂U to ∂V. The point of (3) is that an arbitrary homology h-cobordism is homology h-cobordant rel boundary to one which arises in this way.

Recall that the structure set $S_{\text{TOP}}^h(X)$ is defined to be homology h-cobordism classes of manifolds homology equivalent to X. Parts (1) and (3) of the theorem therefore immediately give:

11.9D Corollary. *Suppose X is a 3-dimensional Poincaré space with good fundamental group. Then the structure set $S_{\text{TOP}}^h(X)$ is in one-to-one correspondence with ends of 4-manifolds with trivial end obstruction, together with homotopy equivalence of the homotopy collar with X.*

Note the equivalence relation on the set of ends is homeomorphism of neighborhoods of the end, so that the resulting homotopy equivalence of homotopy collars homotopy commutes with the given equivalences to X. ∎

This result shows ends can be studied using the surgery machinery of 11.3. A special case will be applied in 11.10.

Proof of the theorem: We begin with (1). The proper deformation of a weak collar defines a map $U \to E$, where E is the homotopy collar, which is a homotopy inverse for $e\colon E \to U$. Applying duality, using the duality proposition for homotopy collars in 11.9A, this implies $\partial U \to U$ is a $\mathbf{Z}\pi_1 U$-homology equivalence. This can also be seen directly: the proper deformation implies the locally finite cohomology $H^*_{lf}(U; \mathbf{Z}\pi)$ vanishes. But this is dual to the relative homology $H_*(U, \partial U; \mathbf{Z}\pi)$. In any case this identifies U up to homotopy as a plus construction (11.2) on ∂U. In particular the homomorphism $\pi_1 \partial U \to \pi_1 U$ is onto and has perfect kernel.

Conversely suppose $\pi_1 N \to \pi$ is onto and has perfect kernel. The manifold plus construction can be applied to $N \times [1, 1+1]$ for all $i \geq 0$ to give 4-manifolds M_i. Let $U = \bigcup_i M_i$ (where boundaries $N \times \{i\}$ in M_{i-1} and M_i are identified). This gives a manifold with boundary N. Each M_i is homotopic to the homotopy plus construction, as are the unions $M_i \cup M_{i+1}$. Therefore $M_i \cup M_{i+1}$ deforms rel $N \times \{i+2\}$ into M_{i+1}. Composing all such deformations gives a proper deformation to the end; the proper map $U \times (0, 1] \to U$ required for the definition of a weak collar. Therefore U is a weak collar which realizes the given data.

We give some detail on the construction of this proper deformation. Define $d_i\colon (\bigcup_{k \geq i} M_k) \times [0, 1] \to U$ by extending the deformation of $M_i \cup M_{i+1}$ into M_{i+1} by the identity on $\bigcup_{k > i+1} M_k$. Let $D_i\colon U \to U$ denote the composition

$$U = \bigcup_{k \geq 0} M_k \xrightarrow{\ d_0(-,1)\ } \bigcup_{k \geq 1} M_k \to \cdots \xrightarrow{\ d_i(-,1)\ } \bigcup_{k \geq 1+1} M_k.$$

Then $d_{i+1}(D_i \times 1)$ is a homotopy from D_i to D_{i+1}. Putting this homotopy on the interval $[\frac{1}{i+1}, \frac{1}{i}]$ defines the required map $U \times (0, 1] \to U$.

Now consider (2). Suppose U is a weak collar. Then according to (1), ∂U together with $\pi_1 \partial U \to \pi_1 U$ is the data for a standard weak collar U_0 constructed as above. We show there is a homeomorphism $U \to U_0$ which is the identity on the boundary. This proves (2), since two collars with the same data will both be homeomorphic to the standard one, therefore homeomorphic to each other.

Since π_1 is good, the proper h-cobordism theorem implies it is sufficient to find a proper s-cobordism from U to U_0. Begin with $U \times [0, \infty)$,
and think of it as a proper cobordism from $U \times \{0\}$ to $\partial U \times [0, \infty)$. Let
W_i denote the $\mathbf{Z}\pi$-homology s-cobordism from $\partial U \times [i, i+1]$ to M_i (notation as in the proof of (1)). Then $\bigcup_{i \geq 0} W_i$ gives a proper $\mathbf{Z}\pi$-homology
h-cobordism from $\partial U \times [0, \infty)$ to U_0. The union $U \times [0, \infty) \cup \bigcup_{i \geq 0} W_i$
is the desired proper s-cobordism.

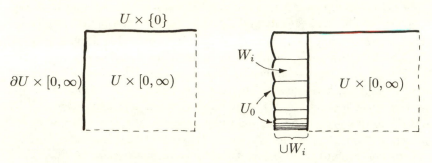

It is a straight-forward matter to verify this is a proper h-cobordism.
To show it is an s-cobordism we actually show the obstruction group is
trivial.

Recall that the proper s-cobordism theorem is derived in 7.3 from
the controlled h-cobordism theorem; a proper h-cobordism with a connected end can be given the structure of a controlled h-cobordism over
$[0, \infty)$. The obstruction group for this is a locally finite homology group
$H_1^{lf}([0, \infty); \mathcal{S}(p))$. $\mathcal{S}(p)$ is the coefficient system resulting from the system of groups with $\pi_1 U$ over $\{0\}$ and $\pi_1(\mathrm{end}U)$ over the rest of $[0, \infty)$.
Since U is a weak collar these groups are the same, and the coefficient
system is constant over $[0, \infty)$. But locally finite homology of $[0, \infty)$
with any constant coefficients is trivial. Thus the obstruction group is
trivial.

We conclude from the proper (or controlled) s-cobordism theorem that
U is homeomorphic to U_0, as required.

Finally consider (3). Suppose U, V are weak collars, $\pi_1 U \simeq \pi_1 V \simeq \pi$
are isomorphisms, and W is a $\mathbf{Z}\pi$-homology h-cobordism from ∂U to
∂V. Specifically this means $\partial W = \partial U \cup \partial V$, there is $\pi_1 W \to \pi$ commuting with the homomorphisms on $\pi_1 \partial U$ and $\pi_1 \partial V$, and with respect
to this homomorphism the relative $\mathbf{Z}\pi$-homology groups of $(W, \partial U)$ and
$(W, \partial V)$ are trivial.

Since π is good and $\dim W = 4$ we can apply the plus construction
to obtain a $\mathbf{Z}\pi$-homology h-cobordism rel boundary of W to $\hat{W}$, with
$\pi_1 \hat{W} = \pi$. Then $\hat{W} \cup_{\partial U} U$ is a weak collar with the same data as V.

According to (2) this means there is a homeomorphism $V \simeq \hat{W} \cup U$. This restricts to a homeomorphism of ends, which has associated h-cobordism $\hat{W}$, as required for the theorem. ∎

11.9E General tame ends. Here we consider the structure and classification of ends which do not have weak collars. The first result is that such ends have a "periodic" structure.

Theorem. *Suppose M is a 4-manifold with compact boundary and connected tame end with homotopy collar E. If $\pi_1 E$ is finitely presented and good, then*

(1) *there is a compact manifold N homotopy equivalent to $E \times S^1$ whose infinite cyclic cover $\hat{N}$ is homeomorphic with a neighborhood of the end of M,*

(2) *there is a homeomorphism $\hat{N} \times S^1 \simeq N \times \mathbf{R}$ commuting up to homotopy with the maps to E, and*

(3) *if N' is a manifold satisfying (1) and (2) for an end proper h-cobordant to the end of M, then N is h-cobordant to N'.*

Before discussing the proof we indicate the connections with "proper surgery." Given an n-dimensional Poincaré space E define the "proper structure set" $S^p_{\text{TOP}}(E)$ to be the subset of $S^h_{\text{TOP}}(E \times S^1)$ represented by N with property (2) in the theorem. Then there is a "proper surgery" sequence

$$L^p_{n+1}(\mathbf{Z}\pi) \to S^p_{\text{TOP}}(E) \to NM_{\text{TOP}}(E) \to L^p_n(\mathbf{Z}\pi)$$

which is exact in the same sense as the ordinary surgery sequences in 11.3. This is essentially the result of Pedersen and Ranicki [**1**]. With this definition we get an analog of 11.9D:

11.9F Corollary. *Suppose E is a 3-dimensional Poincaré space with good fundamental group. Then the proper structure set $S^p_{\text{TOP}}(E)$ is in one-to-one correspondence with proper h-cobordism classes of tame ends of 4-manifolds with a homotopy equivalence of the homotopy collar to E.*

Proof of the theorem: In higher dimensions this is called "furling," and was developed by M. Brown, and L. Siebenmann, using engulfing. In this dimension a different proof is necessary; we use the 5-dimensional end theorem and the proper h-cobordism theorem.

The product $M \times S^1$ has a tame end with vanishing $\tilde{K}_0$ obstruction. Since the fundamental group is good the 5-dimensional version of the collaring theorem (11.9A; Quinn [**4**]) shows that there is a collar $N \times \mathbf{R}$

of the end of $M \times S^1$. This can be used to identify $M \times S^1$ as the interior of a compact manifold $M \times \mathbf{R} \cup N \times \{\infty\}$. This identifies the infinite cyclic cover $\hat{N} \times \mathbf{R}$ as part of a collar for the end of $M \times \mathbf{R}$. Use this to add some boundary to $M \times \mathbf{R}$. Another piece of boundary is obtained by identifying $M \times \mathbf{R}$ with $M \times (0,1)$ and adding $M \times \{0,1\}$. There are still "corners" missing.

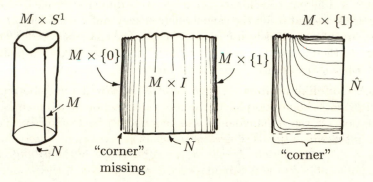

The result can be considered a proper h-cobordism from $M \times \{0\}$ to the cover $\hat{N}$.

According to the proper h-cobordism theorem 9.6 (since π is good) there is an obstruction in $\tilde{K}_0(\mathbf{Z}\pi)$ to the existence of a product structure on this h-cobordism in a neighborhood of the given end of M. We can change this obstruction by attaching an h-cobordism of N to the completion of $M \times S^1$. These h-cobordisms are classified by $Wh(\pi \times \mathbf{Z}) \simeq Wh(\pi) \oplus \tilde{K}_0(\mathbf{Z}\pi) \oplus NIL$, and the $\tilde{K}_0$ summand matches the proper h-cobordism obstruction. Thus there are choices of h-cobordism whose covers yield a proper h-cobordism with trivial invariant. (Note that this modification changes $\hat{N}$.) The resulting product structure gives a homeomorphism of the end of M with a cyclic cover, as required.

For property (2), note the end of $M \times S^1$ has collars $N \times \mathbf{R}$ by choice of N, and $\hat{N} \times S^1$ by the proof above. These 5-dimensional collars are unique up to homeomorphism, so $N \times \mathbf{R} \simeq \hat{N} \times S^1$.

Finally (3) is proved by applying a relative 5-dimensional version of the proof above to the h-cobordism; we omit details. ∎

11.10 Neighborhoods of fixed points

We discuss the structure of neighborhoods of fixed points in actions on 4-manifolds. In most cases this can be shown to be the same as actions on $\mathbf{R}^4$ which fix the origin, though we will not use this. The emphasis is on actions which are free in the complement of isolated fixed points.

Fixed points will be divided into four classes of increasing generality; linear, conelike, weakly conelike (which have compact contractible invariant manifold neighborhoods), and the general case. In higher dimensions the conelike and weakly conelike cases coincide, essentially because the plus construction can be applied to the boundary of the neighborhood. They separate in this dimension because the boundary is 3-dimensional. However a different coincidence occurs; in this dimension conelike and linear are expected to be the same, whereas they differ considerably in higher dimensions. This is discussed in 11.9A. The analysis of ends using surgery in the last section gives "programs" for classifying fixed points in the other two cases.

11.10A Conelike actions. A fixed point is *conelike* if a neighborhood is equivariantly homeomorphic to the cone on an action on a 3-manifold. The action on the 3-manifold is also required to be locally conelike, which in this dimension is equivalent to smoothness. Briefly, most—and probably all—such actions are linear; a neighborhood of the fixed point is homeomorphic to a real 4-dimensional representation of the group.

Note that the 3-manifold in the cone must be a homotopy sphere. It is conjectured that every smooth action of a finite group on a homotopy 3-sphere is "essentially linear" in the sense that the action is obtained from a linear action on S^3 by adding copies of a homotopy sphere at points in a free orbit, and extending the action to permute these. This has been announced to be true for actions which are not free by Thurston [1], extending earlier work by many people which established the cyclic case (the Smith conjecture; Morgan and Bass [1]). For free actions it is known when the order of the group is divisible only by the primes 2 and 3, and there are encouraging signs in other cases (Thomas [1]).

The next result shows the 3-dimensional conjecture implies that conelike fixed points are linear.

Lemma. *The open cone on an essentially linear action on a homotopy 3-sphere is homeomorphic to the open cone on a linear action.*

Proof: Suppose an action of G on M^3 is essentially linear. Identifying the homotopy cells to points gives an equivariant map to the linear action on S^3. This map is cell-like and a homeomorphism in the complement of the preimage of the free orbit where the homotopy spheres were added. Multiply by $\mathbf{R}$ to get a cell-like map of 4-manifolds. The quotient $(M \times \mathbf{R})/G \to (S^3 \times \mathbf{R})/G$ is a homeomorphism in the complement of the preimage of an arc $\{x\} \times \mathbf{R}$ which has a manifold neighborhood.

Since cell-like maps between 4-manifolds can be approximated by homeomorphisms, holding fixed closed sets where they are already homeomorphisms (Quinn [4]) the quotient map can be approximated rel a neighborhood of the singular set by a homeomorphism. Using the G cover of this approximation to modify the free part of the product map gives an equivariant homeomorphism $(M \times \mathbf{R})/G \simeq (S^3 \times \mathbf{R})/G$. Adding a point at one end gives a homeomorphism of open cones, as required. ∎

11.10B The relation to ends. We now restrict to actions free in the complement of the fixed point. Some of what follows can be extended to more general actions (along the lines of the exercises in 11.1), but this will not be done here.

The significance of the freeness assumption is that $M/G - x$ is a manifold, with an end where x used to be. There is a homomorphism $\pi_1(M/G - x) \to G$, and M with its action can be recovered from this by taking the G covering space and compactifying by adding a point.

A *Swan space* for a group G is a space with the homotopy type of a CW complex, fundamental group G, and universal cover the homotopy type of some sphere S^k. Since Poincaré duality is preserved by finite coverings, X is a k-dimensional Poincaré space.

Lemma. *Let G be a finite group acting on a manifold M freely in the complement of a point x.*

   (1) *The end of $M/G - x$ is tame, and the homotopy collar is a Swan G-space.*

   (2) *Conversely, if N is a manifold with a tame end whose homotopy collar is a Swan G-space, then the 1-point compactification of the G-cover of a compact invariant neighborhood of the end, is a manifold.*

   (3) *$W \subset M$ is a contractible invariant closed manifold neighborhood of the fixed point if and only if $W/G - x$ is a weak collar for the end of $M/G - x$.*

So, for example, we see that a fixed point has a compact invariant contractible neighborhood if an end obstruction in $\tilde{K}_0(\mathbf{Z}G)$ vanishes.

Proof: The end of $M - x$ is $S^{n-1} \times \mathbf{R}$, so is tame. This is a covering space over the end of $M/G - x$, so since tameness is preserved by coverings this end is also tame. The homotopy collar of the end of $M/G - x$ has as G cover the homotopy collar of the end of $M - x$, so is a Swan G-space.

for (2) we need only verify that the 1-point compactification is a manifold in a neighborhood of the compactification point. The homotopy collar of the complement of this point is S^{n-1} so the space is a homology manifold, and the local fundamental group of the complement is trivial. Therefore the space is a manifold by the flattening theorem for points (9.3A for dimension 4).

Finally consider weak collars. The map in the definition of a weak collar lifts to the covering space to give a contraction of the corresponding neighborhood of x. Conversely suppose W is an invariant contractible manifold neighborhood in M. Since the end of $W/G - x$ is tame there is a proper deformation of some neighborhood to the end. Therefore it is sufficient to find a deformation of $W/G - x$ into an arbitrary neighborhood of the end. For this it is sufficient to show that the homotopy collar $E \rightarrow (W/G - x)$ is a homotopy equivalence. The G cover of this is $S^{n-1} \rightarrow W - x$. The contractibility of W implies this is a homotopy equivalence, and equivalences are preserved by covering spaces so $E \rightarrow (W/G - x)$ is also. Therefore $W/G - x$ is a weak collar. ∎

11.10C Weakly conelike actions. Recall we say a fixed point is "weakly conelike" if it has a compact invariant contractible manifold neighborhood.

According to the lemma above, equivariant homeomorphism classes of neighborhoods of such fixed points correspond exactly to homeomorphism classes of tame weakly collared ends whose homotopy collars are G Swan spaces. For actions on 4-manifolds corollary 11.9C further describes these in terms of surgery: they are determined by homotopy h-cobordism classes of boundaries of weak collars.

Proposition. *There is a one-to-one correspondence between homeomorphism classes of neighborhoods of isolated weakly conelike G fixed points, and* $\bigcup_X S^h_{\text{TOP}}(X)/\sim$.

Here X ranges over the G Swan spaces, and the equivalence relation $\sim$ is generated by homotopy self-equivalences of X. Using the the surgery sequence therefore gives a "surgery program," as in 11.3B, for classifying such fixed points:

(1) classify the 3-dimensional Swan complexes X for G, if any,

(2) determine which of these have topological reductions of the normal fibration, and which reductions (if any) have trivial surgery obstruction in $L^h_3(G)$, and

(3) determine the action of $L^h_4(G)$, and the homotopy self-equivalences of X, on $S^h_{\text{TOP}}(X)$.

This is a large and complex undertaking, with contributions to the higher dimensional analog made by J. Milnor, R. Swan, R. Lee, C. T. C. Wall, C. B. Thomas, I. Madsen, I. Hambleton, R. J. Milgram, and many others. The answers which come out are complicated. At the time of writing the most complete account is Hambleton and Madsen [**1**], which comes close to settling (1) and (2), in terms of number-theoretic conditions on G. (Their calculations for the surgery obstruction $NM \rightarrow L$

apply to dimension 3. This dimension is excluded in the final statements because the surgery sequence was not known to be exact, but section 11.3 repairs this deficiency). We will not describe the conclusions except to say that there are groups which have weakly conelike fixed points which do not have linear ones. Also, there are usually many weakly conelike fixed points not equivalent to linear ones, even when linear ones exist.

We very sketchily illustrate this last point in the case of cyclic groups. Suppose G is cyclic of odd order, and $N = S^3/G$ is a lens space (quotient of a free linear action). Then in the surgery sequence the normal map set is a single point, and $S^h_{\text{TOP}}(N)$ is nonempty since it contains N. The reduced group $\tilde{L}^h_4(G)$ acts freely on S^h so gives a bijection $\tilde{L}^h_4(G) \simeq S^h_{\text{TOP}}(N)$. If G is not trivial then this is an infinite set. But it is a classical result that there are only finitely many lens spaces with a given fundamental group, and there are only finitely many homotopy self-equivalences of each of these. Therefore there are infinitely many nonlinear weakly conelike fixed points for G.

11.10D General actions. As observed above, lemma 11.10B shows the end obstruction of the quotient in $\tilde{K}_0(\mathbf{Z}\pi)$ vanishes if and only if the fixed point is weakly conelike. Theorem 11.9E shows there are equivariantly periodic neighborhoods of the fixed point. Corollary 11.9F gives a surgery formulation of a classification of these fixed points, which we discuss next.

Say that isolated fixed points of G actions on M and M' are *concordant* if there is a G action on a manifold with boundary W which is free in the complement of a locally flat fixed arc, and ∂W is the union of neighborhoods of the fixed points in M and M'. It is easily seen as in 11.10B that concordances W correspond exactly to proper h-cobordisms of the ends of $M/G - x$ and $M'/G - x$. Therefore 11.9F gives:

Proposition. *There is a one-to-one correspondence between concordance classes of neighborhoods of isolated G fixed points, and*

$$\bigcup_X S^p_{\text{TOP}}(X)/ \sim .$$

Here, as in 11.10C, X varies over the G Swan spaces and the equivalence relation $\sim$ is generated by self-equivalences of the X. Again this gives a "surgery program" for approaching the classifcation of such fixed points:

   (1) classify the 3-dimensional Swan complexes X for G, if any,
   (2) determine which of these have topological reductions of the normal fibration, and which reductions (if any) have trivial surgery obstruction in $L^p_3(G)$, and

(3) determine the action of $L_4^p(G)$ on $S_{\mathrm{TOP}}^p(X)$, etc.

Some of this program has been carried out in higher dimensions by Hambleton and Madsen [2]. As above the methods now apply to the 4-dimensional case. The results themselves are rather complicated, so we only remark that there are groups G which have isolated semifree fixed points, but cannot have compact neighborhoods as in the previous section.

CHAPTER 12

Links, and Reformulations of the Embedding Problem

The embedding problem concerns the necessity of fundamental group restrictions in the disk embedding theorem. The reformulations given in this chapter were developed to explore the problem; either suggest new approaches or bring invariants to bear which would detect counterexamples.

The first section restates the problem in terms of embedding disks in neighborhoods of properly immersed capped gropes. These neighborhoods (when orientable) occur as subsets of D^4; the second section describes these subsets explicitly as complements of certain link slices. The embedding problem is therefore equivalent to a link slice problem.

The third section concerns slices for links in S^3. The links arising in the embedding problem are shown to be "good boundary links," so surgery predicts the existence of slices. In fact, surgery is equivalent to the link slice problems, and a weak form of the embedding problem. Special subclasses of links which have been shown to be slice are described in 12.3D.

Section 12.4 gives a Poincaré transversality criterion for the existence of manifolds equivalent to Poincaré complexes with free fundamental group. Since these can be used to construct slices for links, this gives another form of the embedding problem.

The final section describes a reformulation in terms of actions of free groups on S^3.

12.1 The embedding problem

The main embedding result for 2-disks is stated in the introduction, and in Theorem 5.1. The data are an immersed 2-disk, and a framed immersed 2-sphere with algebraically trivial selfintersections and algebraic intersection 1 with the disk. The conclusion is that if the fundamental group is poly-(finite or cyclic) then there is an embedded disk with the same framed boundary. The embedding "problem" is to decide if the statement is true without the fundamental group hypothesis. The embedding problem "up to s-cobordism" asks if there is a disk embedded in a manifold s-cobordant (rel boundary) to the original. The most interesting results concern this weaker problem.

Note if we take a regular neighborhood of the data then we get an embedding problem in a much more specific manifold, and the general problem is equivalent to such restricted ones. Capped gropes will be used for this rather than a disk and transverse sphere, because the regular neighborhoods are simpler.

Lemma. *The embedding problem is equivalent to the existence of embedded disks with the given framed boundary in (orientable) neighborhoods of properly immersed 2-stage capped gropes. The embedding problem up to s-cobordism is equivalent to embedding in grope neighborhoods up to s-cobordism.*

Proof: The proof of the embedding theorem produces properly immersed capped gropes from the immersed disk and sphere data, so the grope version implies the general case. Conversely, a new capped grope can be found in a neighborhood of a given one so that the cap intersections appear in pairs with immersed Whitney disks which have immersed transverse spheres. Applying the disk-and-sphere version to the Whitney disks gives embedded Whitney disks. These can be used to get embedded caps for the grope. But then the disk embedded in a neighborhood of the model capped grope has image an embedded disk in in the manifold.

These capped gropes can be arranged to have orientable neighborhoods even if the manifold is not orientable; use 2.9 to get a grope with the property that the image has fundamental group in the kernel of $\omega_1\colon \pi_1 M \to \mathbf{Z}/2$. ∎

We remark that this approach to the problem looses contact with the fundamental group of the manifold. Thus it does not give a way to expand the class of "good" groups without solving the whole problem. Also we expect embeddings near gropes of height 1.5, since the heights of these can be raised arbitrarily (see the exercise in 2.7).

12.2 Link pictures

The objective is to draw link pictures describing neighborhoods of properly immersed capped gropes and towers. As in Part I the approach is to squeeze as much as possible of a 4-dimensional situation into a 3-dimensional slice (the "present"). This method, developed by the second author, is complementary to the "moving picture" descriptions used by R. Kirby and his school. The basic data for such a picture is a link $b \cup L \subset S^3$ such that b is a single unknotted circle, and L is a trivial link (bounds disjointly embedded disks). To this data we associate a manifold $M_L \subset D^4$ by pushing the interiors of 2-disks spanning L radially a short way into the interior of D^4, and deleting open tubular neighborhoods from D^4. The circle b determines a framed embedded loop in ∂M_L.

The first point is that these manifolds are well defined. The disks spanning L are not included in the data, but a standard 3-manifold theorem asserts they are nearly well-defined up to isotopy rel boundary. Different collections of disks differ by sums with boundaries of 3-cells containing the components of L. When radially deformation into the interior these become isotopic, so the complement is well defined up to isotopy.

The next point is that these manifolds have standard handlebody structures, with one 0-handle and some orientable 1-handles. Deletion of the tubular neighborhoods is the same as addition of a 1-handle; the radial deformation applied to D^2 gives a copy of $D^2 \times I$ in M_L, which is a 3-handle in $(M_L, \partial M_L)$ dual to a 1-handle.

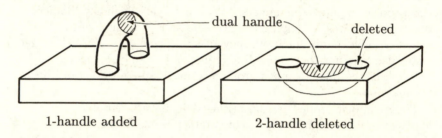

1-handle added   2-handle deleted

Replacing the tubular neighborhoods to get D^4 back corresponds to adding 2-handles to cancel the new 1-handles. The 2-disks in the tubular neighborhood are dual to these cancelling 2-handles. In particular L itself consists of boundaries of duals of 2-handles used to cancel the new 1-handles.

This notation can be extended to describe handlebodies which also have 2-handles and—with some difficulty—3-handles; see Kirby, Kas, and Harer [1], and Kirby [1], [2]. This extension will not be needed here.

12.2A Neighborhoods of surfaces. We give link descriptions of neighborhoods of bodies of capped surfaces. Following the conventions of Part I (see section 2.1) a capped surface with genus 1 is drawn in the "present" $D^3 \times \{0\} \subset D^3 \times [-1, 1]$. The body is obtained by deleting the 2-handle caps, so the link picture will consist of b, the boundary of the surface, and L the boundaries of 2-disks dual to the caps. These duals are drawn as arcs, using the convention that arcs are extended to surfaces by crossing with the $[-1, 1]$ factor.

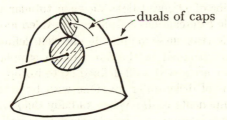

Regard S^3 as the double of D^3, one copy representing the "past," the other the "future." To facilitate drawing this double, we draw the picture so that the present intersections with the boundary lie in a flat 2-disk, then double along this 2-disk.

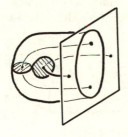
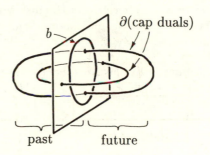

Isotoping the boundaries of the cap duals into a standard solid torus linking b shows they are obtained by replacing a linking circle for b with the Bing link.

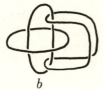

A capped surface with several handles can be obtained by boundary connected sum of several copies of the single-handle case. The link picture is therefore obtained by Bing doubling several copies of the linking circle of b. Generally we define a *ramified Bing double* of a framed link to be obtained by taking parallel copies of each component, and then replacing each of these by the Bing link. Note this denotes a class of links rather than a specific one because we have not specified how many parallel copies of each component to use. Using this terminology a neighborhood of the body of a capped surface is M_L, where L is a ramified Bing double of the linking circle of the boundary curve b.

12.2B Composition of link pictures. Suppose $b \cup L$ and $b' \cup L'$ are links as above, and $c \in L$ is a component. Compose by compressing L' into a solid torus linking b', and then applying an isomorphism of this solid torus with a neighborhood of c. There is a unique way (up to isotopy) to do this so that the boundary of the normal disk of b' is identified with the boundary of the disk bounding c. Note the link $L - c \cup L'$ is trivial; disks bounding L' can be compressed into a neighborhood of the normal disk to b', so the images in the composed link bound disks in a neighborhood of the disk bounding c. Since L is trivial, such a neighborhood can be chosen disjoint from disks bounding the other components of L.

Geometrically this corresponds to taking a union $M_L \cup M_{L'}$ and identifying a neighborhood of $b' \subset \partial M_{L'}$ with the $S^1 \times D^2 \subset M_L$ used to attach the 2-handle to cancel c. In other words, we replace the 2-handle with $M_{L'}$.

Capped gropes provide the first examples. A capped grope is defined by replacing the caps of a capped surface with other capped surfaces. The link picture for the body of the result is therefore obtained by composing the picture for the original surface with pictures for the cap replacements. Note this gives a ramified Bing double of the original link.

Define a *j-fold iterated (ramified) Bing double* of a framed link to be a link obtained by performing j ramified Bing doublings. With this notation, we see that neighborhoods of bodies of capped gropes with j stages are represented by j-fold iterated Bing doubles of the linking circle of the boundary curve b.

12.2C Disks with intersections. Consider a disk with a single (orientable) selfintersection. Adding an accessory disk (see 3.1) gives a 2-complex with neighborhood D^4, so a neighborhood of the disk with selfintersection is obtained by deleting the accessory disk. As above the link we want is the boundary of the dual of the accessory disk.

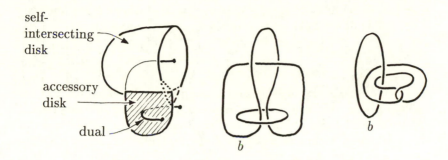

The picture identifies L as obtained by replacing the linking circle of b with a Whitehead link. Note there are two of these, with clasps differing by rotation by 1/2 turn, corresponding to selfintersections with sign $+1$ or -1. Again disks with multiple selfintersections can be obtained by connected sum of multiple copies of single ones, so have links obtained by replacing multiple copies of the linking circle of b with Whitehead links.

Define a *ramified Whitehead double* of a framed link to be obtained by taking multiple copies of each component, then replacing each of these by a Whitehead link. In this notation, ramified Whitehead doubles of the linking circle of b give link pictures for neighborhoods of disks with multiple selfintersections.

Now we compose this with the picture for grope bodies to get link pictures of immersed capped gropes. Suppose a capped grope with j stages is immersed in an orientable 4-manifold so that the only intersections are selfintersections in individual caps. Then a neighborhood of the image is M_L, where L is obtained by a ramified Whitehead doubling of a j-fold iterated Bing doubling of the linking circle of the boundary curve b.

12.2D Whitney disks. The final ingredient for pictures of neighborhoods of capped towers is a picture of a Whitney, accessory disk pair (see 3.1). proceeding as above we get

Whitney disk

accessory disk

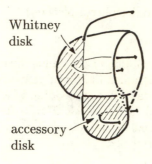
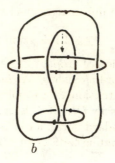
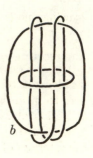

which when compressed into a linking solid torus gives a 2-component link we call the Whitney link;

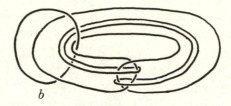

Define a *ramified Whitney double* of a framed link to be a link obtained by replacing parallel copies of the link by copies of the Whitney link. A neighborhood of a disk with several pairs of intersection points is therefore represented by M_L, where L is obtained by a ramified Whitney double of a linking circle of the boundary curve b.

We recall that a 1-story capped tower (with grope height j) is obtained by replacing the caps of a capped grope of height j with selfintersecting disks with Whitney and accessory disks. Suppose such a tower is immersed in an orientable 4-manifold so that the only intersections are selfintersections of the Whitney and accessory disks. Then a neighborhood of the image is given by M_L, where L is a ramified Whitehead double, of a ramified Whitney double, of a j-fold Bing double, of the linking circle of b.

We have found Whitney disks convenient in the manipulation of towers, but it is possible to use two accessory disks in place of a Whitney, accessory pair. This simplifies the link picture a little, by replacing the Whitney double with a Whitehead double. With this modification a neighborhood of a tower is obtained by a 2-fold Whitehead double of a j-fold Bing double of the linking circle.

Exercise Draw an example of a link describing a neighborhood of a properly immersed capped tower. (These are complicated enough that drawing one is more informative than trying to decipher a picture.) □

12.3 Link slice problems

The link description of gropes gives a reformulation of the embedding problems in terms of link slicing. A *slice* for a framed link in S^3 is used here to mean an extension to a topological locally flat embedding of a collection of 2-disks in D^4. If L is the trivial link (bounds 2-disks in S^3), then the "standard slice" for L is obtained by radially deforming the interiors of bounding 2-disks into the interior of D^4.

12.3A Proposition. *Consider the class of links $b \cup L$, where b is an unknotted circle and L is a ramified Whitehead double of a j-fold iterated Bing double of the linking circle of b, $j \geq 2$.*

(1) *The existence of slices for $b \cup L$ whose restriction to L is the standard slice, is equivalent to the embedding problem for disks in manifolds with arbitrary fundamental group.*

(2) *The existence of slices for $b \cup L$ whose restriction to L has free complementary fundamental group, is equivalent to the embedding problem up to s-cobordism for arbitrary fundamental groups.*

This type of formulation was pioneered by Casson [1], and predates any solutions to the embedding problem. A significant amount of the material of Part I was discovered by studying these links; see eg. Freedman [2], [4]. There is also a well-developed theory of link invariants showing some links to be unsliceable, which may provide an approach to a negative solution to the embedding problems. The links in the proposition are immune to all invariants discovered to date.

Proof: We use the grope formulation given in 12.1. According to the previous section a neighborhood of the image of a properly immersed 2-stage capped grope is isomorphic to M_L, where L is a link in the specified class. A slice for b in the complement of the standard slices for L is exactly an embedded disk in M_L, so a solution to the embedding problem.

Conversely, we have constructed 2-stage properly immersed capped gropes in the complement of the standard slice for L, with b as framed boundary. If the embedding problem has an affirmative solution then there is an embedded disk in a neighborhood of this grope, therefore a slice for $b \cup L$.

Now we consider the s-cobordism statement. A slice for L with free complementary fundamental group has complement homotopy equivalent to a 1-complex, by 11.6C(1). This means the complement is s-cobordant to the standard slice complement, by 11.6A. Therefore a slice for b in this complement is an embedding up to s-cobordism in the standard complement.

Conversely, suppose there is an embedding up to s-cobordism in the standard complement. Attaching 2-handles to fill in where the slices were removed gives a slice for $b \cup L$, in a contractible manifold with boundary S^3, so that the slice for L has free complementary fundamental group. The implication is completed by observing the contractible manifold must be D^4, by the Poincaré conjecture. ∎

12.3B Good boundary links. In this section the links encountered above are shown to be good boundary links, which means there is a homomorphism of the complement to a free group, which takes linking circles to generators and which has perfect kernel (see 11.7C).

Lemma. *Suppose L_0 is a link with all linking numbers trivial, F is an orientable surface with a single boundary component b, and F is embedded disjoint from L_0. Suppose there is another embedding F' of the surface so that linking numbers between elements of L_0 and curves on F are the same as with the corresponding curves on F'. Suppose also that $b' = \partial F'$ bounds a disk disjoint from L_0. If L is a ramified Whitehead double of L_0 then $b \cup L$ is a good boundary link.*

The "good boundary" condition is equivalent to a condition on linking numbers of curves on Seifert surfaces. Comparison with a trivial link avoids the need to spell the condition out, or to calculate to verify it.

Proof: Suppose $F_1, \dots, F_n$ are disjoint oriented surfaces in S^3, each with a single boundary component f_i. (In the application of this discussion these will be Seifert surfaces for the Whitehead double of L_0.) The orientations determine an orientation of the normal line bundles of these surfaces, so give parallel copies lying in the complement. Define $p_j^+, p_j^- : F_j \to S^3 - \cup F_i$ to be parallels on the positive and negative sides respectively.

Alexander duality gives an isomorphism between $H_1(S^3 - \cup f_i)$ and the dual $\left(\sum_i H_1(F_i)\right)^*$. The Seifert linking form is the composition of this with the positive parallel $p^+ : \sum_i H_1(F_i) \to H_1(S^3 - \cup F_i)$. Explicitly this is given by linking numbers; a curve c on the surface f_j is taken to the homomorphism which assigns to a curve d the linking number of $p_j^+(c)$ and d. Choosing curves representing a basis for $\sum_i H_1(F_i)$ gives a representation of this form by a matrix of such linking numbers.

Using the negative parallels of the surfaces gives the dual of this form, which is represented by the transpose matrix.

Denote the free group on n generators $x_1, \dots, x_n$ by $\mathcal{F}^n$. A homomorphism $\pi_1(S^3 - \cup f_i) \to \mathcal{F}^n$ can be defined by intersections with the F_i. If a loop is transverse to the surfaces then at an intersection point with F_j associate x_j or x_j^{-1} depending on the sign of the intersection. Associate a word in the free group by forming the product of elements encountered traveling around the loop. The kernel of this homomorphism is perfect if the first homology of the corresponding cover of $S^3 - \cup f_i$ is trivial. We therefore describe this cover.

Let M denote $S^3 - \cup f_i$ cut open along the surfaces. This has boundary two copies of the interior of each surface, which we identify with the positive and negative parallels; $\partial M = \bigcup_i (p_i^+(\text{int} F_i) \cup p_i^-(\text{int} F_i))$. The cover is given by $\bigcup_{g \in \mathcal{F}^n} gM$, with identifications $g p_i^-(\text{int} F_i) \sim (g x_i) p_i^+(\text{int} F_i)$. The Mayer-Vietoris sequence for the homology of this cover shows H_1 depends only on the homomorphisms $H_1(p_*^\pm)$. These in turn are determined by the linking form.

The conclusion is that if two embeddings of the surfaces F_* in S^3 have the same linking forms, then the kernels of $\pi_1(S^3 - \cup f_i) \to \mathcal{F}^n$ have isomorphic abelianizations. In particular if the boundary of one of the embeddings is the trivial link, then the boundary of the other is a good boundary link.

Now let $b \cup L_0$ be the link of the lemma, and L a ramified Whitehead double of L_0. We reduce first to the case where each component of L_0 is replaced by a single Whitehead link: the hypotheses on L_0 are preserved if parallel copies of components are added, so replace L_0 by the multiple used in the ramified double. The next picture shows a Seifert surface for the Whitehead link, with curves x, y which give a basis for H_1.

Now replace each component c_i of L_0 with a Whitehead link, with Seifert surface F_i containing curves x_i, y_i. The key point is to note that the linking numbers of x_i and y_i with any curve disjoint from the original c_i is the same as the linking number of that curve with c_i. This means the linking form on the $\cup F_i$ is determined by the linking numbers with L_0.

The first application of this is to discard b. Since the two surfaces F and F' have the same linking numbers with L_0, the surfaces $F \cup \bigcup_i F_i$ and $F' \cup \bigcup_i F_i$ have the same linking forms. Therefore it is sufficient to verify $b' \cup L$ is a good boundary link. But b' was presumed to bound a disk disjoint from L, so since the condition depends on the link and not on the surfaces chosen, we can replace F by a disk. It follows it is sufficient to verify L is a good boundary link.

Next, since the linking numbers between components of L_0 are the same as the trivial link, the Whitehead double of L_0 has the same linking form as the Whitehead double of the trivial link. But the double of the trivial link is trivial, so is a good boundary link. Therefore so is $b \cup L$. $\blacksquare$

12.3C Examples of good boundary links. Suppose L_0 is any link with trivial linking numbers. Then a Whitehead double of L_0 is a good boundary link. This follows from the lemma by letting b be the boundary of a disk disjoint form L_0. The simplest example is a Whitehead double of the Borromean rings:

It seems likely that the slice problem for this link is the crucial test case for the entire problem.

Now suppose b is unknotted, and L_1 is any link disjoint from b. Construct L_0 by (ramified) Bing doubling L_1, and then further Bing doubling one component of each Bing pair (a 1.5-fold Bing double). We claim this L_0 satisfies the hypotheses of the lemma. To see this, arrange after the first Bing double that only one of each Bing pair intersects the disk bounding b. A Seifert surface for b disjoint from this link can be constructed by surgery on the disk, along the link. Bing double the components which intersected the disk, then the components of the result all have trivial linking numbers with curves in the Seifert surface. These are the same linking numbers as would have been obtained if the curves had been pushed through the Seifert surface (and the disk) to remove the linking with b, before doubling. Thinking of this as giving a different embedding of the Seifert surface gives a situation satisfying the hypothesis of the lemma.

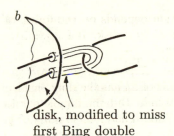

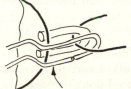

disk, modified to miss   second Bing double, made disjoint
first Bing double   from surface and disk

We note that Whitehead doubling a 1.5-fold iterated Bing double of the linking circle of b gives the picture of a properly immersed capped grope of height 1.5. According to the exercise in 2.7 this height is exactly what is needed to raise the height of the grope, so the problem in some sense is already in a "stable range."

Combining this with 11.7C and 12.3A shows:

Corollary. *"Surgery" is equivalent to embedding up to s-cobordism, and to the existence of "free slices" for good boundary links.*

We note that surgery and the s-cobordism theorems together are equivalent to the full embedding problem, and to existence and uniqueness of free slices. (Compare the second part of corollary 11.7C with the role of the standard slices in 12.3A(1).)

12.3D Links known to be slice. Good boundary links with one component (ie. knots with Alexander polynomial 1) are slice, according to theorem 11.6C. A Whitehead double of a 2 component link with trivial linking numbers is slice, by Freedman [9]. As suggested above, the 3-component case is likely to be equivalent to the general problem.

A boundary link has all linking numbers trivial, so Whitehead doubles of these are good boundary links. These also have been shown to be slice, by Freedman [5]. Unfortunately Bing doubles are not boundary links, so this does not apply to the embedding problem. Note that the embedding theorem for disks in a capped tower (or the π_1-null theorem 5.2) asserts the existence of slices after two Whitehead doublings. Since a Whitehead double is a boundary link, this is a special case of the boundary link result.

12.3E The A-B slice problem. In Freedman [8] the "A-B slice" condition is defined. Roughly, a link L_0 is A-B slice if and only if Whitehead doubles of it are free-slice. The problem thus becomes to determine if the Borromean rings, for example, are A-B slice. Calculations of Freedman and Lin [1] show that manifolds appearing in an A-B slice structure must have a certain minimum homological complexity, but as yet no general results have been obtained.

12.4 Poincaré transversality

In this section the link slice problem, therefore the embedding problem up to s-cobordism, is shown to be equivalent to a "homotopy theoretic problem" of dividing up CW complexes satisfying Poincaré duality. The result comes from Freedman [6], and uses the π_1-null criterion for embedding up to s-cobordism derived in Chapter 6. Comments and definitions follow the statement.

Theorem. *Suppose (X, N) is a Poincaré pair with N a 3-manifold, and $f \colon X \to \vee^k S^1$ an isomorphism on π_1 whose restriction to N is transverse to $k \subset \vee^k S^1$. Then X is homotopy equivalent to a manifold with boundary N if and only if there is a manifold normal map rel N to (X, N), and f can be made $\mathbf{Z}$-Poincaré transverse to k, rel N.*

In the context of slices for good boundary links the normal maps exist (11.6B, 11.7C). Therefore the existence of slices (and so the embedding problem up to s-cobordism) is equivalent to the Poincaré transversality problem. None of the current techniques for proving Poincaré transversality apply in this dimension (Jones [1], Quinn [10], Hausmann and Vogel [1]), nor have any obstructions been identified which might show transversality fails.

In the link slice situation the transversality corresponds to Poincaré "Seifert surfaces" for slices. A good boundary link $L \subset S^3$ is the boundary of a collection of Seifert surfaces $S \subset S^3$ obtained as the transverse inverse image $f^{-1}(k)$, where $f \colon (S^3 - L) \to \vee^k S^1$. If L extends to "free slices" $K \subset D^4$ then the map extends to $F \colon (D^4 - K) \to \vee^k S^1$. The inverse image $F^{-1}(k)$ then is a collection of 3-manifolds which are "Seifert surfaces" for the slices; these have boundary $S \cup K$. For a general good boundary link we can construct a Poincaré slice: a Poincaré embedding of 2-disks in a space homotopy equivalent to D^4 whose boundaries are the given link (11.7C). The theorem implies that if we can make the complement of this Poincaré embedding transverse to $k \subset \vee^k S^1$, to obtain "Poincaré Seifert surfaces" for these slices, then we can find genuine topological slices.

Definition. We define "$\mathbf{Z}$-Poincaré transverse." $k \subset \vee^k S^1$ denotes a collection of points, one in each circle, with collar neighborhood $k \times [-1, 1]$. The manifold N is transverse to k if $N \cap f^{-1}(k \times [-1, 1])$ is of the form $(N \cap f^{-1}(k \times \{0\})) \times [-1, 1]$, and the $[-1, 1]$ coordinate of f on this is the projection. Poincaré transversality is defined similarly, except that X is allowed to change up to homotopy. Thus f is Poincaré transverse rel N if there is a homotopic pair $(\hat{X}, N)$ and homotopic map $\hat{f}$ which is equal to f on N, so that $\hat{f}^{-1}(k \times [-1,1])$ is of the form $\hat{f}^{-1}(k \times \{0\}) \times [-1,1]$, the $[-1,1]$ coordinate of the map is the projection, and $\hat{f}^{-1}(k \times \{0\})$ is a **Z**-Poincaré pair. We also require that the inverse image of k and its complement have finitely presented fundamental groups.

For the purposes here we define a pair (X, Y) to be n-dimensional **Z**-Poincaré if there is a "fundamental class" $[X] \in H_n(X, Y; \mathbf{Z})$ so that the cap products $\cap [X] \colon H^i(X) \to H_{n-i}(X, Y)$ and $\cap [X] \colon H^i(X, Y) \to H_{n-i}(X)$ are isomorphisms. It follows from this and the exact sequences of the pair that $\partial[X] \in H_{n-1}(Y)$ is a fundamental class, so Y is **Z**-Poincaré of dimension $n - 1$. Similarly exact sequences show that the "complement" in the definition of transversality is **Z** Poincaré, namely $\left(\hat{X} - \hat{f}^{-1}(k \times (-1,1)), (N - \hat{f}^{-1}(k \times (-1,1))) \cup \hat{f}^{-1}(k \times \{-1,1\}) \right)$. See Browder [**1**, §1] for results of this sort.

The fundamental group of a **Z**-Poincaré space is automatically finitely generated; it seems to be an additional hypothesis to require it to be finitely presented.

Proof of the theorem: First note that if (X, N) is equivalent rel N to a manifold, then this manifold provides a normal map. Also the manifold transversality theorem implies it can be made topologically, thus Poincaré, transverse to $k \subset \vee^k S^1$.

For the other implication we suppose (by changing it up to homotopy) that $f \colon X \to \vee^k S^1$ splits in the manner described in the definition. Let $g \colon M \to X$ be the normal map. Modify this to have trivial surgery obstruction using the action of $L_4(\text{free}) = L_4(\{1\})$ on the normal map set (see the reduced surgery sequence in 11.3B, or the proof of 11.6A).

Make g transverse to $f^{-1}(k)$, and arrange $g^{-1}(f^{-1}(k)) \to f^{-1}(k)$ to be a **Z**-homology isomorphism. For this it is sufficient to show the map restricted to the inverse image is normally bordant to a homology equivalence, since such a restricted normal bordism can be extended to a normal bordism of g. However the existence of such a normal bordism of the inverse image follows from a slight modification of the proof that simply-connected surgery is unobstructed in odd dimensions (Browder [**1**, IV.3]). Make the map an isomorphism on π_0, then the fact that $\pi_1(*) \to H_1(*; \mathbf{Z})$ is onto replaces all references to the Hurewicz theorem, and avoids the need for simple connectivity. The homology computations go through without change.

Let $(Y, \partial Y)$ denote the complement $X - f^{-1}(k) \times (-1,1)$, as a Poincaré pair. Let $(W, \partial W) \to (Y, \partial Y)$ denote the corresponding complement in

M. Using the finite presentability of the fundamental group of Y we can do 0- and 1-surgeries on W so that $W \to Y$ induces an isomorphism on π_1.

Since g has been arranged to be a homology equivalence over $f^{-1}(k)$, the map of boundaries $\partial W \to \partial Y$ is a homology equivalence. This, and the vanishing of the surgery obstruction of g imply that the relative homology $H_i(Y, W; \mathbf{Z})$ vanishes for $i \neq 3$, and in that dimension is free with a basis $\{a_i, b_i\}$ so that the intersection form on the image in $H_2(W)$ is represented by a hyperbolic matrix $\begin{bmatrix} 0 & I \\ I & 0 \end{bmatrix}$. The next step is to represent the images of this basis in $H_2(W)$ by maps of 2-spheres into W.

The exact sequence of Whitehead [1] for W and Y gives a diagram

$$
\begin{array}{ccccccccc}
\pi_2 W & \longrightarrow & H_2 W & \longrightarrow & \Gamma_2 W & \longrightarrow & \pi_1 W & \longrightarrow & H_1 W \\
\downarrow & & \downarrow & & \downarrow & & \downarrow & & \downarrow \\
\pi_2 Y & \longrightarrow & H_2 Y & \longrightarrow & \Gamma_2 Y & \longrightarrow & \pi_1 Y & \longrightarrow & H_1 Y
\end{array}
$$

The two right vertical homomorphisms have been arranged to be isomorphisms. Whitehead shows $\Gamma_2 \simeq H_2(\pi_1; \mathbf{Z})$ and therefore depends only on π_1. Thus the middle vertical is also an isomorphism. This implies the kernel of $H_2 W \to H_2 Y$ is in the image of $\pi_2 W$. Let $\hat{a}_i, \hat{b}_i \colon S^2 \to W$ be maps representing the classes a_i, b_i obtained above.

Now we return to $M \to X$. This map is 2-connected, and

$$
H_3(X, M; \mathbf{Z}[\pi_1 X]) \simeq H_3(Y, W; \mathbf{Z}) \otimes_{\mathbf{Z}} \mathbf{Z}[\pi_1 X]
$$

is an isomorphism of $\mathbf{Z}[\pi_1 X]$-modules. Further, the intersection form is induced from the one on $H_3(Y, W; \mathbf{Z})$. This implies that if we can represent the maps $a_i \colon S^2 \to M$ by disjoint π_1-negligible embeddings, then surgery on them will produce a manifold homotopy equivalent to X.

The union of the images of the a_i and b_i lie in W, which goes trivially into $\pi_1 X \simeq \pi_1 \vee^k S^1$. These spheres therefore satisfy the hypotheses of Theorem 5.3, which asserts that M is s-cobordant to a manifold in which they can be embedded. Specifically, the a_* are homotopic in the new manifold to disjoint framed embeddings, with embedded transverse spheres homotopic to the b_*. The conditions for successful surgery are homotopy invariant, so surgery on these embeddings yields the manifold equivalent to X, as required. ∎

12.5 Actions of free groups

Let $L \subset S^3$ denote a ramified Whitehead double of a j-fold Bing double of the Hopf link, and suppose it has k components. According to 12.3B this is a good boundary link; let $\pi_1(S^3 - L) \rightarrow \mathcal{F}^k$ be the homomorphism to the free group. Then $\mathcal{F}^k$ acts freely on the associated cover of $S^3 - L$. In Freedman [7] the endpoint compactification (see 3.8) of this cover is shown to be S^3. Since endpoint compactification is functorial, the action of the free group extends to an action on S^3. The limit set (where the action is not free) is the set of endpoints, thus a Cantor set.

If L is free-slice then the endpoint compactification of the free cover of closed complement of the slices is D^4. The action of the free group on S^3 therefore extends to an action on D^4, still free in the complement of the Cantor set in the boundary. The theorem is that this process is reversible: slices can be reconstructed from appropriate extensions of the $\mathcal{F}^k$ action on S^3 to an action on D^4. Consequently this extension problem is equivalent to the embedding problem up to s-cobordism. We refer to Freedman [7] for the proof, and precise definitions of the classes of actions considered.

These actions are analogous to "Schottky group" actions on $(n - 1)$-spheres. Extensions for these actions arise naturally by considering the conformal geometry on S^{n-1} as the "boundary" of the hyperbolic geometry on $\mathbf{R}^n$. This analogy has been studied by Freedman and Skora [1], [2], but the extension problem equivalent to the embedding problem remains unresolved.

References

F. D. Ancel
1. *Approximating cell-like maps of S^4 by homeomorphisms*, in "Four-manifold Theory," Gordon and Kirby ed., Contemporary Math. **35** (1984), 143–164.

F. D. Ancel and M. P. Starbird
1. *The shrinkability of Bing-Whitehead decompositions*, To appear in Topology.

J. J. Andrews and M. L. Curtis
1. *Free groups and handlebodies*, Proc. AMS **16** (1965), 192–195.

J. J. Andrews and L. Rubin
1. *Some spaces whose product with E^1 is E^4*, Bull. Amer. Math Soc. **71** (1965), 675–677.

R. J. Bean
1. *Decompositions of E^3 with a null sequence of starlike equivalent nondegenerate elements are E^3*, Illinois J. Math **11** (1967), 21–23.

R. H. Bing
1. *A homeomorphism between the 3-sphere and the sum of two solid horned spheres*, Ann. Math. **56** (1952), 354–362.
2. *Upper semicontinuous decompositions of E^3*, Ann. Math **65** (1957), 363–374.
3. "The geometric topology of 3-manifolds," Amer. Math Soc., Providence, RI, 1983.

S. Boyer
1. *Simply connected 4-manifolds with a given boundary*, Trans. Amer. Math. soc. **298** (1986), 331–357.

W. Browder
1. "Surgery on simply-connected manifolds," Springer-Verlag, New York, 1972.
2. *Structures on $M \times R$*, Proc. Cambridge Phil. Soc. **61** (1965), 337–345.

J. L. Bryant and C. L. Seebeck III
1. *Locally nice embeddings in codimension three*, Quart. J. Math. Oxford **21** (1970), 265–272.

J. W. Cannon
1. *The recognition problem: what is a topological manifold?*, Bull. AMS **84** (1978), 832–866.

2. *Shrinking cell-like decompositions of manifolds. Codimension 3*, Ann. Math.
   110 (1979), 83–112.

J. W. Cannon, J. L. Bryant, and R. C. Lacher
1. *The structure of generalized manifolds having nonmanifold set of trivial dimension*, in "Geometric Topology, J. C. Cantrell ed," Academic Press, 1979, pp.
   261–300.

S. E. Cappell and J. L. Shaneson
1. *The codimension two placement problem and homology equivalent manifolds*,
   Ann. Math. **99** (1974), 277–348.
2. *On four-dimensional s-cobordisms*, J. Diff. Geometry **22** (1985), 97–115.
3. *On four-dimensional s-cobordisms, II*, To appear.

A. Casson
1. *Lectures on new infinite constructions in 4-dimensional manifolds*, notes by L.
   Guillou, Orsay.

A. Casson and M. H. Freedman
1. *Atomic surgery problems*, AMS Contemporary Math **35** (1984), 181–200.

A. Casson and C. Gordon
1. *On slice knots in dimension three*, in "Proceedings of Symp. in Pure Math,
   Vol. 32 (Algebraic and geometric topology)," Amer. Math. Soc., Providence
   RI, 1978.

T. A. Chapman
1. "Controlled simple homotopy theory and applications," Lecture notes in math-
   ematics 1009, Springer-Verlag, 1983.
2. *Locally homotopically unknotted embeddings in codimension 2 are locally flat.*

T. D. Cochran and N. Habegger
1. *On the homotopy theory of simply-connected four-manifolds.*

M. M. Cohen
1. "A course in simple-homotopy theory," Springer-Verlag, New York, 1970.

R. Connelly
1. *A new proof of M. Brown's collaring theorem*, Proc. Amer. Math Soc. **27** (1971),
   180–182.

R. J. Davermann
1. *Sewings of n-cell complements*, Trans. Am. Math Soc.
2. "Decompositions of manifolds," Pure and Appl. Math vol. 124, Academic Press,
   1986.

S. K. Donaldson
1. *An application of gauge theory to the topology of 4-manifolds*, J. Diff. Geom.
   18 (1983), 279–315.
2. *Connections, cohomology, and the intersection forms of 4-manifolds*, J. Diff.
   Geom. **24** (1986), 275.
3. *The orientation of Yang-Mills moduli spaces and 4-manifold topology*, J. Diff.
   Geom. **26** (1987), 397–428.
4. *Irrationality and the h-cobordism conjecture*, J. Diff. Geom. **126** (1987), 141.

B. Eckmann and P. Linnell

1. *Poincaré duality groups of dimension two, II*, Comment. Math. Helvetici **58** (1983), 111-114.

R. D. Edwards

1. *The solution of the 4-dimensional Annulus conjecture (after Frank Quinn)*, in "Four-manifold Theory," Gordon and Kirby ed., Contemporary Math. **35** (1984), 211–264.

R. D. Edwards and R. C. Kirby

1. *Deformations of spaces of embeddings*, Ann. Math. **93** (1971), 63–88.

F. T. Farrell and W-C. Hsiang

1. *The Whitehead groups of poly-(finite or cyclic) groups*, J. London Math. Soc. **24** (1981), 308–324.
2. *The topological characterization of flat and almost flat Riemannian manifolds M^n, (n ≠ 3, 4)*, Am. J. Math. **105** (1983), 172–180.

F. T. Farrell and L. E. Jones

1. *The surgery L-groups of poly-(finite or cyclic) groups*, Invent. Math. **91** (1988), 559–586.
2. *Topological rigidity for Hyperbolic manifolds*, preprint.

S. C. Ferry

1. *Homotoping ϵ maps to homeomorphisms*, Am. J. Math **101** (1979), 567–582.

R. Fintushel and R. J. Stern

1. *Pseudofree orbifolds*, Ann. Math. **122** (1985), 335–364.
2. *SO(3) connections and the topology of 4-manifolds*, J. Diff. Geometry **20** (1984), 523–539.

D. S. Freed and K. Uhlenbeck

1. "Instantons and four-manifolds," MSRI publications, Springer-Verlag, New York, 1984.

M. H. Freedman

1. *A fake $S^3 \times \mathbf{R}$*, Ann. Math. **110** (1979), 177–201.
2. *The topology of four-dimensional manifolds*, J. Diff. Geom. **17** (1982), 357–453.
3. *A surgery sequence in dimension four: the relations with knot concordance*, Invent. Math. **68** (1982), 195–226.
4. *The disk theorem for four dimensional manifolds*, in "Proc. Int. Congress Math. Warsaw 1983," pp. 647-663.
5. *A new technique for the link slice problem*, Invent. Math. **80** (1985), 453–465.
6. *Poincare transversality and four-dimensional surgery*, Topology **27** (1988), 171–175.
7. *A geometric reformulation of 4-dimensional surgery*, Topology and Appl. **24** (1986), 133–141.
8. *Are the Borromean rings A-B slice?*, Topology and Appl. **24** (1986), 143–145.
9. *Whitehead₃ is slice*, Inventiones Math..

M. H. Freedman and R. Kirby

1. *A geometric proof of Rochlin's theorem*, Proc. Symp. Pure Math. AMS **32 II** (1978), 85–98.

M. H. Freedman and X. S. Lin
  1. *On the A − B slice problem*, Topology **28** (1989), 91–110.

M. H. Freedman and F. Quinn
  1. *A quick proof of the 4-dimensional stable surgery theorem*, Comment. Math.
     Helv. **55** (1980), 668–671.
  2. *Slightly singular 4-manifolds*, Topology **20** (1981), 161–173.

M. H. Freedman and R. Skora
  1. *Strange actions of groups on spheres*, J. Diff. Geom. **25** (1987), 75–98.

M. H. Freedman and L. R. Taylor
  1. *A universal smoothing of four-space*, J. Diff. Geometry **24** (1986), 69–78.

H. Freudenthal
  1. *Neuaufbau der endentheorie*, Ann. Math **43** (1942), 261–279.

R. Friedman and J. Morgan
  1. *On the diffeomorphism type of certain algebraic surfaces*, J. Diff. Geometry **27**
     (1988), 297–369.
  2. *Algebraic surfaces and 4-manifolds: Some conjectures and speculations*, Bull.
     Amer. Math. Soc. **18** (1988), 1–19.

R. E. Gompf
  1. *Three exotic $\mathbf{R}^4 s$ and other anomalies*, J. Diff. Geom. **18** (1983), 317–328.
  2. *An infinite set of exotic $\mathbf{R}^4 s$*, J. Diff. Geom. **21** (1985), 283–300.
  3. *A moduli space of exotic $\mathbf{R}^4 s$*.

R. E. Gompf and S. Singh
  1. *On Freedman's reimbedding theorems*, in "Four-manifold Theory," Gordon and
     Kirby ed., Contemporary Math. **35** (1984), 277–310.

V. Guillemin and A. Pollack
  1. "Differential topology," Prentice Hall, Englewood Cliffs NJ., 1974.

L. Guillou and A. Marin
  1. "A la Recherche de la Topology Perdue," Progress in Math. 62, Birkhäuser,
     Boston, 1986.

N. Habegger
  1. *Une variété de dimension 4 avec forme d'intersection paire et signature−8*,
     Comment. Math. Helvetici **57** (1982), 22–24.

I. Hambleton and M. Kreck
  1. *On the classification of 4-manifolds with finite fundamental group*, Math. Ann
     280 (1988), 85–104.

I. Hambleton and I. Madsen
  1. *Local surgery obstructions and space forms*, Math. Zeitschrift **193** (1986),
     191–214.
  2. *Actions of finite groups on $\mathbf{R}^{n+k}$ with fixed set $\mathbf{R}^k$*, Can. J. Math. **38** (1986),
     781–860.

J. Harer, A. Kas, and R. Kirby
  1. *Handlebody decompositions of complex surfaces*, Memoirs Amer. Math. Soc. **62**
     (1986).

J-C. Hausmann and P. Vogel
  1. "Poincaré spaces" (to appear).

M. Hirsh and B. Mazur
  1. "Smoothings of piecewise-linear manifolds," Annals of Math Studies, Princeton
    University Press, 1974.

F. Hirzebruch and W. Neumann and S. Koh
  1. "Differentiable manifolds and quadratic forms," Lecture notes vol. 4, Marcel
    Dekker, 1971.

L. S. Hush and T. M. Price
  1. *Finding a boundary for a 3-manifold*, Ann. Math. **91** (1970), 223–235.

L. Jones
  1. *Patch spaces*, Ann. Math. **97** (1973), 306–343.

O. Kakimizu
  1. *Finding boundary for the semistable ends of 3-manifolds*, Hiroshima Math. J.
    17 (1987), 395–403.

M. A. Kervaire and J. W. Milnor
  1. *On 2-spheres in 4-manifolds*, Proc. Nat. Acad. USA **47** (1961), 1651–1657.

R. C. Kirby
  1. *A calculus for framed links in S^3*, Invent. Math. **45** (1978), 35–56.
  2. "The topology of 4-manifolds," To appear.

R. C. Kirby and L. C. Siebenmann
  1. "Foundational essays on topological manifolds, smoothings, and triangula-
    tions," Annals of Math Studies, Princeton University Press, 1977.
  2. *Normal bundles for codimension 2 locally flat embeddings*, in "Park City Topol-
    ogy conference proceedings", Springer Lecture Notes in Math. **438** (1975),
    310–324.

K. Kobayashi
  1. *On a homotopy version of 4-dimensional Whitney's lemma*, Math Seminar
    Kobe Univ. **5** (1977), 109–116.

S. Kwasik and R. Schultz
  1. *Non-locally smoothable topological symmetries of 4-manifolds*, Duke J. Math.
    55 (1986), 765–770.

H. B. Lawson Jr.
  1. "Theory of gauge fields in four dimensions," CBMS monograph, Amer. Math.
    Soc., Providence, 1985.

R. Lashof and L. Taylor
  1. *Smoothing theory and Freedman's work on four manifolds*, in "Algebraic Topol-
    ogy Aarhus 1982," Springer lecture notes, Springer-Verlag, 1984, pp. 271–292.

A. Marin
  1. *La transversalite topologique*, Ann. Math. **106** (1977), 269–293.

J. W. Milnor and D. Husemoller
  1. "Symmetric bilinear forms," Springer-Verlag, 1973.

J. W. Milnor
    1. *Microbundles, part I*, Topology **3** (1964), 53–80.
    2. *Whitehead torsion*, Bull. Amer. Math. Soc. **72** (1966), 358–426.
    3. "Lectures on the h-cobordism theorem," Princeton University Press, 1965.

J. W. Morgan and H. Bass
    1. "The Smith conjecture," Pure and Appl. Math vol. 124, Academic Press, 1984.

R. A. Norman
    1. *Dehn's lemma for certain 4-manifolds*, Invent. Math. **7** (1969), 143–147.

E. K. Pedersen and A. Ranicki
    1. *Projective surgery theory*, Topology **19** (1980), 239–254.

R. Penrose, J. H. C. Whitehead, and E. C. Zeeman
    1. *Imbedding of manifolds in euclidean space*, Ann. Math. **73** (1961), 613–623.

D. Quillen
    1. *Cohomology of groups*, in "Actes, Congrès Int. Math. 1970, Tome 2," pp. 47–51.

F. Quinn
    1. *The stable topology of 4-manifolds*, Top. and Appl. **15** (1983), 71–77.
    2. *Ends of maps, I*, Ann. Math. **110** (1979), 275–331.
    3. *Ends of maps, II*, Invent. Math. **68** (1982), 353–424.
    4. *Ends of maps, III: dimensions 4 and 5*, J. Diff. Geom. **17** (1982), 503–521.
    5. *Embedding towers in 4-manifolds*, in "Four-manifold theory," R. Kirby and C. Gordon ed., AMS Contemporary Math **35** (1984), 461–472.
    6. *Smooth structures on 4-manifolds*, in "Four-manifold theory," R. Kirby and C. Gordon ed., AMS Contemporary Math **35** (1984), 473–479.
    7. *Geometric algebra*, in "Algebraic and Geometric Topology," A. Ranicki et al. ed., Springer lecture notes **1126** (1985), 182–198.
    8. *Isotopy of 4-manifolds*, J. Diff. Geom. **24** (1986), 343–372.
    9. *Topological transversality holds in all dimensions*, Bull. Amer. Math. Soc. **18** (1988), 145–148.
    10. *Surgery on Poincaré and normal spaces*, Bull. Amer. Math. Soc. **78** (1972), 262–267.
    11. *Homotopically stratified sets*, J. Am. Math. Soc. **1** (1988), 441–499.

A. Ranicki
    1. *The algebraic theory of surgery*, Proc. London Math Soc. **40** (1980), 87–283.

C. P. Rourke and B. J. Sanderson
    1. "Introduction to piece-wise linear topology," Ergebnisse der Math., Springer-Verlag, New York, 1972.
    2. *Block bundles*, Ann. Math. **87** (1968), 1–28.

L. C. Siebenmann
    1. *Infinite simple homotopy types*, Indag. Math. **32** (1970), 479–495.
    2. *Regular open neighborhoods*, General topology **3** (1973), 51–61.

S. Smale
    1. *Generalized Poincare's conjecture in dimensions greater than 4*, Ann. Math. **64** (1956), 339–405.

M. Scharlemann
  1. *Transversality theories at dimension 4*, Invent. Math. **33** (1976), 1–14.

M. A. Stan'ko
  1. *Approximations of imbeddings of compacta in codimension greater than two*, Dokl. Akad. Nauk. SSSR **198** (1971); translation Soviet Math. Dokl. **12** (1971), 906–909.

C. Taubes
  1. *Gauge theory on asymptotically periodic 4-manifolds*, J. Diff. Geom. **25** (1987), 363–430.

R. L. Taylor
  1. *Relative Rochlin invariants*, Topology and its Applications **18** (1984), 259–280.

C. B. Thomas
  1. "Elliptic structures on 3-manifolds," London Math. Soc. Lecture series, Cambridge Univ. Press, Cambridge, 1986.

W. P. Thurston
  1. *Three-manifolds with symmetry*, (preprint).

P. Vogel
  1. *On simply connected 4-manifolds with boundary*, in "Algebraic topology Aarhus 1981," Aarhus Univ., Aarhus, 1982, pp. 116–119.

C. T. C. Wall
  1. "Surgery on Compact Manifolds," Academic press, New York, 1970.
  2. *Geometrical connectivity I*, J. London Math Soc. **3** (1971), 579–604.
  3. *Finiteness conditions for CW complexes, I*, Ann. Math **81** (1965), 56–69.

J. H. C. Whitehead
  1. *Note on the previous paper entitled 'On adding relations to homotopy groups'*, Ann. Math. **47** (1946), 806–810.

H. Whitney
  1. *The self-intersections of a smooth n-manifold in 2n-space*, Ann. Math. **45** (1944), 220–246.
  2. *The self-intersections of a smooth n-manifold in 2n − 1-space*, Ann. Math. **45** (1944), 247–293.

Index of Notation

Notations are listed in order of occurrence. References are given by section or subsection rather than page number. Section numbers are printed in the heading on the inside of each page.

$\omega_2(-)$; second Stiefel-Whitney class, 1.3, 1.9.

$\lambda(-,-)$; intersection number, 1.7.

$\mu(-)$; selfintersection number, 1.7.

$\chi(\nu)$; Euler number of ν, 1.7.

$\bar{\omega}$; involution on group ring 1.7

$\#$; connected sum
  of surfaces, 1.8.
  of 4-manifolds, 1.9.

$(-)^t$; transverse sphere, 1.9.

$E(-)$; endpoint compactification, 3.8.

$d(-,-)$; distance between relations, 4.5.

$cl(-)$; closure.

$\Sigma(-)$; singular image of a map, 4.6.

$*M$; M a 4-manifold, 10.1, 10.4.

$e(-)$; ends of $(-)$, 7.3.

$ks(-)$; Kirby-Siebenmann invariant 8.3D, 10.2B.

rE, r-disk bundle of vector bundle E.

$|\lambda|$, 4-manifold with form λ, 10.1.

$roc(-)$; Rochlin invariant of spin 3-manifolds, 10.2B.

$(-)_h$; mapping cylinder of h.

T_+, T_-, 10.5A.

$\#_{S^1}$; S^1-connected sum, 10.7A.

$km(-)$; Kerviare-Milnor invariant, 10.8A.

S_{TOP}; manifold structure set, 11.3.

NM_{TOP}; normal map set, 11.3.

L_n; Wall surgery group, 11.3.

Index of Terminology

Accessory disks, 3.1.

Alexander isotopy, 9.3B.

Algebraic transverse sphere, 1.9.

Almost smoothing, 8.8.

Annulus conjecture, 8.1.

Aspherical, 11.5.

Bing double, 4.2, 12.2A.

Body,
  of a capped surface, 2.1.
  of a capped grope, 2.4.
  of a capped tower, 3.2.

Capped
  surface, 2.1.
  grope, 2.4.
  tower, 3.2.

Compactly supported
  bordism, 11.9B.
  s-cobordism, 5.3, 11.1.

Concordance
  of embeddings, 10.5A
  of group actions, 11.10D.

Conelike (fixed point), 11.10A.

Connected (n-), 8.3C.